techniques of
Observing the Weather

techniques` of

Observing

by

B. C. Haynes
Chief, Observations Section, U. S. Weather Bureau

the Weather

John Wiley & Sons, Inc., New York
Chapman & Hall, Ltd., London

PRINTED IN THE UNITED STATES OF AMERICA

Dedicated to my wife

ANNA LEE

preface

THE beginning of all meteorology, weather forecasting, and climatological studies lies with the original observations of weather and other atmospheric factors. To establish a firm foundation for these sciences a basic and uniform system of observing has been adopted by the International Meteorological Organization. Observing weather is in most countries the responsibility of the government. Observer training has been conducted as an "in-service" program, and all official observing instructions are issued by the government. Most of the official government documents are directed to the experienced observer and are on an "operational" plane.

This book was written for high-school and college courses in elementary meteorology and weather observing. It may also be used by the layman who wants to take up the hobby of weather observing.

Because of the requirement of uniformity in practices of observing, the text follows closely the various United States Weather Bureau instructions on observations.

I wish to express my appreciation to Mr. A. J. Crowshaw for his invaluable assistance in editing the book and to Mr. L. F. Hafer for his assistance in preparing the illustrations. Considerable technical assistance was also given by Mr. L. P. Harrison and Mr. A. V. Carlin.

To Dr. F. W. Reichelderfer, Chief of the United States Weather Bureau, I express gratitude for his interest and encouragement in the preparation of this book.

B. C. HAYNES

Washington, D. C.
November, 1947

contents

illustrations

Chapter I

THE ATMOSPHERE

Definition of Meteorology

Meteorology or, more commonly, weather, is the science, or branch of physics, treating of the atmosphere and its phenomena. There are many kinds of meteors, and they may be classed as follows: (*a*) aerial meteors, such as winds and tornadoes; (*b*) hydrometeors, such as rain, hail, and snow; (*c*) lithometeors, such as dust and smoke; (*d*) luminous meteors, such as rainbows and halos. These also include the igneous meteors, lightning and "shooting stars." Thus the word meteor is not confined to "shooting stars" or meteorites.

Weather Observer

Like chemistry, physics, and other sciences, the study of the atmosphere must begin with known facts. These facts must be determined as accurately as possible and expressed in terms reducible to worldwide standards. The person who makes the original observations of facts concerning the state of the atmosphere is called the *weather observer*. Actually, many elements other than weather may be observed, and *aerographer*, the name used by the United States Navy for this type of observer, seems more appropriate than weather observer. As a science progresses, additional facts are determined. Usually, as new facts are determined, more and more skill is required of the observer, who must be as highly trained as a skilled laboratory technician.

Direct and Indirect Observations

Of all the natural sciences, meteorology comes closest to affecting directly the lives of all people. Even a very young child knows some of the facts about the atmosphere—for example, when it is hot or when it is cold. As we grow older, we observe that clouds obscure the sun and that rain or snow usually occurs when it is cloudy. But such general observations as very hot, very cold, a strong wind, or a great amount of rain have little scientific application. In order to describe the atmosphere with sufficient accuracy and detail for meteorologists to derive the physical laws and make forecasts of weather conditions, a quantitative or numerical description, called an observation, must be made. The usual quantitative method states the ratio of the observed data, in terms of size, weight, or temperature, to some accepted standard. The standard is called the unit. A unit of weight in the English system, for example, is the pound. As each of the elements are taken up in later chapters, the units of measurement will be discussed in detail.

Some observations are made by direct measurements; that is, they are made by comparing the quantity with a unit of the same kind. This is done when we place a ruler in a container of water and measure the depth of the water in inches. In this case, the inch is the accepted standard unit. But most measurements are indirect; for example, the wind speed may be found by measuring the number of miles (units of distance) of wind which pass a fixed point in 1 hour (unit of time) to obtain speed in miles per hour (units of speed). Other observations must be "reduced" or corrected by certain standard procedures to obtain values which may be compared with similarly corrected observations of the same element made at other stations throughout the country or throughout the world.

GENERAL STRUCTURE OF THE ATMOSPHERE [2, 5] *

To understand the scope and magnitude of the observer's work, and some of the variable factors affecting the elements he observes, one must have a knowledge of the structure of the atmosphere.

The atmosphere is principally an ocean of gas which envelops the earth. Air, the basic material composing the atmosphere, is highly elastic and compressible; and, although it is extremely light, it has weight and exerts a pressure on all objects it touches. The weight of unconfined air in a column of unit area that extends

* See numbered references beginning on page 197.

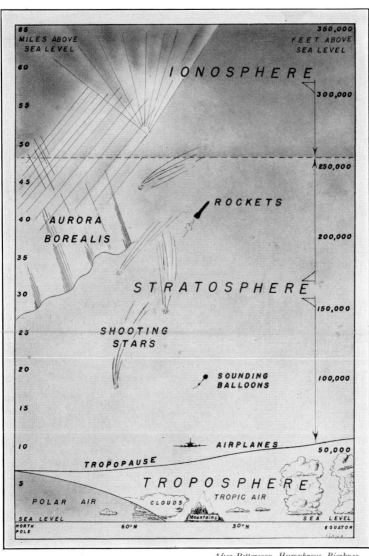

After Petterssen, Humphreys, Bjerknes.

FIG. 1.   Structure of the Atmosphere,

from the point of measurement to the top of the atmosphere is also the pressure of the air at that point. If the point is at sea level, the weight is normally about 15 pounds per square inch. The corresponding units of pressure used in meteorology are discussed in the chapter on pressure. A given volume of air contains 78 per cent nitrogen, 21 per cent oxygen, and about 1 per cent of other gases. The proportions of this gaseous mixture are, for all practical observational purposes, the same in all parts of the world and at all elevations up to at least 12 miles. The atmosphere also contains water, ice clouds, and water vapor, the total varying in amount from 0 to 5 per cent by volume. The atmosphere, even when clear, contains an enormous number of particles or impurities, such as dust particles, smoke particles, and salt particles from sea spray. The density of air continuously decreases with altitude to the upper limit of the atmosphere, the exact height of which is not known. One-half of the atmosphere lies below 18,000 feet (3½ miles), and 90 per cent of the air is contained in a layer only 10 miles deep.[19]

Normally, the temperature decreases upward from the ground, or through the troposphere, to the base of the stratosphere. The dividing line is called the tropopause. Upward through the stratosphere the temperature remains constant or gradually increases. The height of the tropopause varies from day to day, but the average is 55,000 feet over the equatorial regions, and 20,000 feet over the poles.

VARIATIONS IN THE ATMOSPHERE [1, 2]

Most of the time the air is in motion over the greater part of the earth. The direction and speed of movement are expressed as wind velocity. For the most part, the motion is the result of uneven heating and cooling of ocean and land surfaces, which produce the horizontal differences in air temperature. Over the oceans and vegetation-covered land areas, the air absorbs moisture. As the air is cooled by various processes, its moisture, which is usually in the invisible vapor state, condenses out to form fogs and clouds from which rain or snow may fall.

Over the oceans or large land areas of the earth, large, slowly moving bodies of air become warm and moist, or cool and dry. These bodies of air are referred to as air masses. They retain some of their properties as they move away from their source regions, which are usually either polar or tropical. The boundaries separating different, contiguous air masses are called "fronts."

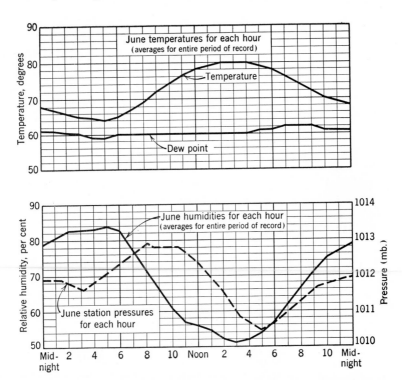

Fig. 2. Diurnal variations of temperature, dew point, relative humidity, and station pressure. Normal curves for Washington, D. C.

They form sloping surfaces that are produced as the lower, colder air mass pushes under the warmer air mass somewhat in the fashion of a flat wedge.

Storm areas in the middle latitudes are the result of interactions between air masses along fronts. These storm areas are usually associated with areas of low pressure, whose wind systems blow counterclockwise around the center in the northern hemisphere, and clockwise in the southern hemisphere. These low-pressure areas, called cyclones, are usually about 1,000 miles in diameter. Their counterparts are areas of high pressure, called anticyclones, about which winds blow clockwise in the northern hemisphere and counterclockwise in the southern hemisphere. The anticyclones are usually areas of good weather, and their diameter averages 2,000 miles.[5, 7]

There are several types of cyclonic storms ranging in size from a diameter of a few hundred yards, as in a tornado, to about 200 miles for the tropical cyclone or hurricane, and up to the cyclone of 1,000 miles in diameter found in the middle latitudes. The cyclonic storms usually move with the general air stream, which is west to east in middle latitudes, and east to west near the equator.

USES OF WEATHER OBSERVATIONS

The foregoing illustrates the scope of the problem which the meteorological observer or aerographer faces. The student who desires to go further into the causes and laws of meteorology should, of course, study textbooks on theoretical meteorology, forecasting, and climatology. With an understanding of the principles of meteorology and a knowledge of the correct *techniques* of weather observing, the student may derive much personal benefit and enjoyment by *observing the weather*.

There are two basic reasons for establishing a weather observatory where regular observations may be made: (*a*) forecasting of future weather conditions; (*b*) study of past conditions to determine climate. Under (*a*) may be listed those observation stations that are established to give up-to-the-minute observations for aviators; under (*b*), all weather observers who keep records are included.

TYPES OF WEATHER STATIONS [18]

The International Congress of Meteorologists, assembled at Vienna in 1873, established the first classification of weather observatories. This classification is still used today.

First-Order Stations of the International Classification

NORMAL METEOROLOGICAL OBSERVATORIES, at which continuous records are kept of hourly readings of pressure, temperature (dry-

and wet-bulb), wind, sunshine, precipitation (rain or snow) with eye-observations of the amount at fixed hours, form and motion of clouds, and notes on the weather.

Second-Order Stations

NORMAL CLIMATOLOGICAL STATIONS, at which are recorded daily, at two fixed hours at least, observations of pressure, temperature (dry- and wet-bulb), wind, cloud, and weather, with the daily maxima and minima of temperatures and the daily amount of precipitation and remarks on weather. At some stations the duration of bright sunshine is also recorded.

Third-Order Stations

AUXILIARY CLIMATOLOGICAL STATIONS, at which observations of the same kind as those at the normal climatological stations are taken, but are (1) less full, or (2) taken once a day only, or (3) taken at other than recognized hours.

In addition to the above stations, there are two classes of stations which measure conditions in the upper air. These are:

Winds-aloft stations, which obtain the direction and velocity of the wind at standard levels up to above 20,000 feet.

Upper-air sounding stations, which obtain temperature, pressure, and humidity up to 70,000 feet or higher.

STANDARD OBSERVATION TIMES

In order that weather maps may be prepared, synchronous observations are made throughout the country and world. Standard observation times for four daily surface observations have been chosen. These times for United States zones are:

1	2	3	4
1:30 A.M. EST	7:30 A.M. EST	1:30 P.M. EST	7:30 P.M. EST
12:30 A.M. CST	6:30 A.M. CST	12:30 P.M. CST	6:30 P.M. CST
11:30 P.M. MST	5:30 A.M. MST	11:30 A.M. MST	5:30 P.M. MST
10:30 P.M. PST	4:30 A.M. PST	10:30 A.M. PST	4:30 P.M. PST

In addition to the above, some stations take intermediate, 3-hourly observations, and hourly observations are taken between the hours specified above.

The full observation is usually made within the 20-minute period just prior to the time specified as observation time.

Upper-air observations are made either twice or four times a day. In the United States these times are:

1	2	3	4
5:00 A.M. EST	11:00 A.M. EST	5:00 P.M. EST	11:00 P.M. EST

In meteorology it is convenient to use the 24-hour clock, 0000 being midnight, or the beginning of the new day. Then 0600 would be 6:00 A.M., and 1800 would be 6:00 P.M.; 2400 is not used to denote midnight.

It is desirable also to refer all observation times to the meridian of Greenwich, England, which is on the zero meridian. This time is referred to as Greenwich Civil Time, G.C.T. For the United States the four time zones may be referred to G.C.T. as follows:

1. Eastern Standard Time—add 5 hours to obtain G.C.T.
2. Central Standard Time—add 6 hours to obtain G.C.T.
3. Mountain Standard Time—add 7 hours to obtain G.C.T.
4. Pacific Standard Time—add 8 hours to obtain G.C.T.

ELEMENTS AND METHODS USED IN WEATHER OBSERVING

The general scope and nature of the problem confronting the meteorological observer have been indicated. Let us now be more specific about the elements that must be observed to determine the facts concerning the state of our ever-changing atmosphere. There are two general classes of surface observations, that is, observations that may be made by an observer on the ground without the aid of the usual sounding devices. Together with instruments or aids used, these may be listed as follows: [10]

Visual Observations

OBSERVATION OF

1. Clouds (types and amount)	By comparison with *International Atlas* types, and by eye
2. Visibility	By eye, objects at known distances
3. Weather	Definitions of hydrometeors

Instrument Observations

OBSERVATION OF	INSTRUMENTS
1. Cloud direction	Nephoscope
2. Cloud height	Ceiling light projector and clinometer Ceiling balloon Ceilometer
3. Temperature	Thermometer Thermograph
4. Humidity	Wet- and dry-bulb psychrometer Hygrometer Hygrograph
5. Wind direction	Wind vane
6. Wind speed	Anemometer
7. Pressure	Mercurial barometer Aneroid barometer Barograph
8. Precipitation Snow depth	Rain gage Measuring stick

In addition to the above surface observations, upper-air observations may be made by ground observers by means of sounding balloons, which are usually free balloons sent high into the air.

The first type of free balloon observation is for obtaining the direction and speed of the winds aloft. These are called pilot-balloon observations, and the instruments used are:

1. A rubber balloon, filled to a given free lift with helium or hydrogen (the ascensional rate is assumed to be constant).

2. A theodolite, similar to a surveyor's transit, through the telescope of which the balloon may be observed.

3. A clockwork time signaler.

4. A radar direction device, if a metallic target is attached to the balloon.

The second type of free balloon observation is the radiosonde observation. A radiosonde is a meteorograph and radio transmitter used in conjunction with a ground radio receiver and recorder. The elements measured are (1) pressure, (2) temperature, (3) humidity. In the following chapters, techniques for observing each of the above elements will be fully discussed.

Chapter II

THE WEATHER OBSERVATORY

SELECTION OF THE SITE

Meteorological observations are of greatest use when they are made under conditions comparable with those of observations made at all other locations. They must therefore be made according to uniform procedures and with accurate instruments. But merely to satisfy this essential condition is not sufficient. Variations of exposure and location of instruments often produce a greater difference in observations than minor inaccuracies in instruments. A surface observation should measure the true conditions that represent the state of the atmosphere in the vicinity of the observatory. Heat from buildings may influence air temperature readings. Wind and precipitation measurements are affected by proximity to buildings.[10, 18]

Selection of a site and the proper exposure of instruments are the first steps in the establishment of the weather observatory. Selection of a site may depend upon the reasons for setting up an observation station. If observations are required at an airport and the observer must be located in or near an administration building, the choice of site will, of course, be limited. If a physics or science department of a school wishes to establish an observatory, the location of instruments indoors and outdoors will be limited to the facilities and grounds at the disposal of the school.

There may be occasions to set up observatories for special agricultural purposes, as for fruit-frost work, at points that would be considered undesirable for general climatic or synoptic purposes.

If the observation station is set up to obtain the most representative measurements of the state of the atmosphere near the ground, then the site chosen for exposing the outside instruments should be the most unrestricted area possible. An open space 300 or 400 feet square without buildings or trees and without large areas of concrete bordering the site (as in the case of an airport) would afford the ideal location.

To prevent the direct rays of the sun from falling on the thermometers used for air temperature measurements, a screen, or

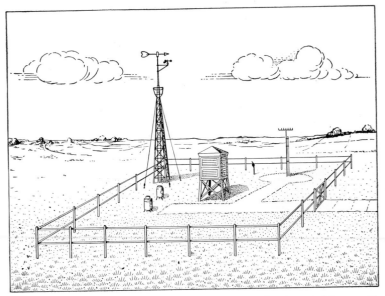

FIG. 3. Typical plot for meteorological instruments showing wind tower with anemometer and wind vane, instrument shelter, rain gages, and comb nephoscope.

instrument shelter, is required. It should be placed near the center of a plot covered by short, level grass, with a minimum area of about 20 feet by 20 feet. The grass cover reduces reflection of radiation and does not become locally heated as a concrete surface might. The rain gage should be placed about 10 feet from the shelter. The plot should be in an area of level ground. A steep slope or a sheltered hollow where air drainage would make the observation not representative should be avoided.

There should also be space inside a building for the observers to work and for the installation of the interior instruments. Normally, a room 10 feet by 10 feet will suffice if the work is to be confined to observing. Classroom demonstration observations may require multiple sets of instruments so that a number of student observers may make nearly simultaneous observations.

LOCATION AND ELEVATION OF OBSERVATORY AND POSITION OF INSTRUMENTS

Climatic studies require detailed knowledge of the exact location of the observation point. The location of the observatory should be given as follows:

> Number, street, city, county, and state; and latitude and longitude to degrees and minutes.

There are a number of elevations which must be established in any observatory in order to make the observations comparable with those made at other stations. A standard official "reference plane" may be determined by using a bench mark established by the United States Coast and Geodetic Survey, the United States Geological Survey, the Lake Survey, the Mississippi River Commission, or the Engineer Corps. The city or county engineer, or one of the above agencies, will be able to locate a convenient bench mark. The elevation of the bench mark in feet and tenths or hundredths of feet is referred to as the "reference plane" elevation.

A line of levels should be run from the bench mark to the building in which the observatory is located, and a secondary bench mark, or "fixed point," should be established on the stonework of the building. Further elevations may then be determined by reference to the "fixed point." The elevations to be established are:

(a) Actual elevation of barometer (H_z) is the elevation of the ivory point or zero point of the barometer scale.

(b) Official elevation of station (H) is the height of the ground above sea level at the station.

(c) Field elevation (H_f) is the elevation above sea level of the average of all possible landing points at an airport.

(d) At airports, elevation of 8-foot plane above field (H_8) is the elevation above sea level of the plane 8 feet above "field elevation" (H_f). This is the average height of altimeters in airplanes on the ground at the individual airport.

(e) Station elevation (H_b) is the elevation above sea level adopted for a station as the basis to which all pressure observations at the station are correlated. For stations located on airports, it is convenient to use the "elevation of the 8-foot plane" as the "station elevation." Some older observatories refer to the "height of the barometer" as "station elevation." Once this elevation is fixed, it should remain the same for reference purposes, even though the barometer may be moved. If the barometer is moved, a *removal correction* is usually applied to the pressure readings in order that the original reduction tables may continue to be used.

ORIENTATION AND SPECIFICATION OF DIRECTIONS

In meteorology all directions are based upon true north. A number of instruments, including the wind vane and nephoscope

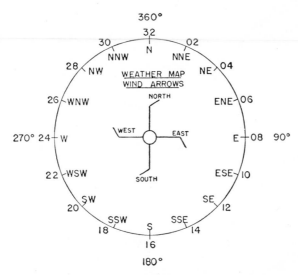

Fig. 4.   Azimuth directions from point of observation.

for observing cloud direction, must be oriented correctly with respect to directions. Direction is specified by cardinal points of

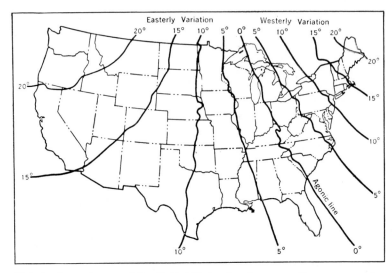

FIG. 5. Isogonic map of the United States showing magnetic compass variation (declination) from true geographical directions.

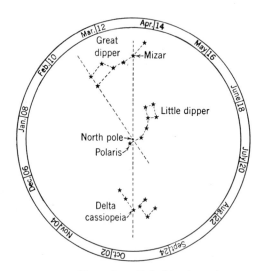

FIG. 6. North Star (Polaris) orientation.

the compass, or degrees from true north measured in a clockwise direction. Figure 4 shows the two methods of measuring direction.

The direction of an object with reference to the observation point is called its "bearing." The bearings of a number of prominent objects, such as towers, spires, or hills, may be determined for reference purposes. True north may be established by referring to a standard survey map. If this method is used, the observation point should be accurately located on the map and bearings to prominent features should be noted. From these bearings, a true north point may be located.

A second method of determining north is with a pocket or surveyor's compass. A magnetic compass reads magnetic north and must be corrected for declination since the magnetic pole is not the same as the geographic pole. Airway maps contain lines (isogonic lines) showing magnetic declination that may be used to correct the compass readings.

A third method is based upon the position of the Pole Star. In northern latitudes the North Star, or Polaris, may be located easily on any clear night as the last star in the "tail" of the constellation known as the Little Bear (Ursa Minor). It may also be located by following a straight line projected across the two bright stars farthest from the tail of the Great Bear or "big dipper." By sighting on the North Star and then dropping vertically down to the horizon, a point close to true north may be located.

WEATHER RECORD FORMS

The weather observatory should have on hand a stock of forms upon which to record observations. A well-kept and unbroken record of observations is often an invaluable asset when climatic studies are made of the area in which the observatory is located. Several types of record forms are suggested, including the forms now in use by the Weather Bureau, Army, and Navy.[10]

RECORD OF LOCATION AND ELEVATION OF OBSERVATORY AND POSITION OF INSTRUMENTS

At _____

| Number | Street | City | County | State |

Latitude _____ Longitude _____

Date of establishment of observatory _____

Date of first observation _____ Time _____

The points from which measurements are made will be accurately described, so that they can be readily found

Authority for measurements	Feet and decimals

Height of barometer (ivory point) above some *fixed point* at base of building

in which office is located _____

Description of fixed point _____

_____

Fixed point _____ reference plane _____

_____

Barometer above reference plane _____

Reference plane above mean sea level at _____

Description of reference plane —————————

Actual elevation of barometer (ivory point) above mean sea level, from the

above figures, H_z = —————

Actual elevation of barometer ——— its actual elevation in old office ——— feet

Actual elevation of barometer ——— H_b, the "station elevation" ——— feet

"Station Elevation," H_b = ————— ft.; "Official Elevation of Station" (Ground), H = ————— ft.

"Field Elevation," H_f = ————— ft.; "Elevation of 8-foot plane above field," H_8 = ————— ft.

Is the shelter roof or sod? ————— Is the shelter standard pattern, cotton region, or special construction? —————

Its inside measurements are ——— ft. long, ——— ft. wide, ——— ft. high, and its floor is ——— ft. above roof or sod.

Height of dry-bulb thermometer above roof rh_t = ———; above ground h_t = ———; ——— barometer —————

   " top of rain gage   " rh_r = ———; " h_r = ———; " —————

   " anemometer cups   " rh_a = ———; " h_a = ———; " —————

   " wind vane   " rh_w = ———; " h_w = ———; " —————

DESCRIPTION OF TOPOGRAPHY AND EXPOSURE OF INSTRUMENTS AT WEATHER REPORTING STATIONS

Station and state _____ County _____ Date _____

Date established _____ Direct supervising station _____

Type of station. First order: _____ Second order: _____

EXPOSURE AND INSTALLATION OF INSTRUMENTS

Instrument shelter: Height of floor above roof ____ ft., above ground ____ ft., height of dry bulb above shelter floor ____ ft. Lighted from: ( ) 110-v. source, ( ) other. Forms used: Thermograph, No. _____ Hygrograph, No. _____ Hygrothermograph, No. _____ Explain briefly any local influences which might affect temperature values. Evaluate the character of exposure as excellent, good, fair, or not representative.

Wind instruments: Anemometer cup-wheel ____ ft. above roof, ____ ft. above ground; wind vane ____ ft. above roof, ____ ft. above ground. Cable ____ ft., number of conductors ____, gage ____, ( ) lead covered, ( ) braid. Describe wiring and installation; give direction and distance of nearby objects obstructing the free movement of wind. Evaluate exposure. If direct-reading dial wind indicators are in use, describe the type and indicate ownership.

Pressure-measuring instruments: Elevation of ivory point above sea level _____ ft.

Mercurial, station: Made by _____ Type _____ Scale _____

Mercurial, extra: Made by _____ Type _____ Scale _____

Barograph: Made by _____ Scale _____ ( ) 4-day, ( ) 7-day. Uses Form No. _____

Aneroid: Made by _____ Scale _____ Describe fully any special

SURFACE WEATHER OBSERVATIONS
(LAND STATION)

Station _____
Lat. _____ Long. _____

Time entries on this form are _____ th meridian.
To convert { add } _____ hours
to G. C. T. { subtract }

Height of Barometer _____ Ft. (MSL)

TYPE (1)	TIME (2)	CLAS-SIFICA-TION (3)	CEILING (hundreds of feet) (4)	SKY (5)	VISI-BILITY (Miles) (6)	WEATHER AND/OR OBSTRUCTIONS TO VISION (7)	SEA-LEVEL PRESSURE (mbs.) (8)	TEMP. AND DEW PT. (°F.) (9) (10)	WIND DIREC-TION (11)	VELOC-ITY (12)	CHAR-ACTER AND A-V-A SHIFTS (13)	ALTIM-ETER (inches) (14)	REMARKS AND SUPPLEMENTAL CODED DATA (For Transmission) (15)	PRES. TEND. (16)	NET 3-HR. CHANGE (mbs.) (17)	PRECIP. (18)	CLOUDS LOW AMT. TYPE, DIRECTION (19)	HEIGHT (mbs. or ft.) (20)	MIDDLE AMT. TYPE, DIRECTION (21)	HEIGHT (mbs. or ft.) (22)	HIGH AMT. TYPE, DIRECTION (23)	HEIGHT (mbs. or ft.) (24)	STATION PRESSURE (inches) (25)	DRY BULB (°F.) (26)	WET BULB (°F.) (27)	WATER TEMP. (°F.) (28)	OBS. INIT. (30)

SUMMARY OF DAY (midnight to midnight)

Time of sunrise _____ Time of sunset _____
24-hr. Max. Temp. _____ °F. 24-hr. Min. Temp. _____ °F.
24-hr. Precip. (rain equiv.) _____ in. 24-hr. Snowfall (snowfall) _____ in.
Total hours of fog: Light _____ Moderate _____ Heavy _____
Max. wind (5 min.) _____ (m.p.h., knots) Direction _____ Time _____
Extreme wind (inst'l inst mis.) _____ (m. p. h., knots) Direction _____ Time _____
Peak gust _____ (m.p.h., knots) Direction _____ Time _____
Thickness of ice on water _____ in.
Ground frozen from _____ in. to _____ in.

6-HOURLY OBSERVATIONS

	Time of Obs.	Midnight
	Precip. (in.)	
	Snowfall (in.)	
	Snow Depth (in.)	
	Max. Temp.	
	Min. Temp.	
	1,000 ft. Pressure	
	Rel. Hum. %	
	Soil Temp. State of Ground	
	State of Sea and Dir.	
	Ht. and Dir. Swell	
	Swell Period	
	Surf (La.L.K.P.D)	
	Barograph Reading	
	Attached Thermometer	
	Observed Reading	
	Total Correction	

Remarks, Notes, and Miscellaneous Phenomena

Types of air masses and frontal passages determined from synoptic
maps, severe storms, miscellaneous hydrometeors, floods

Month _____ Day _____ Year _____

21

Chapter III

CLOUDS [9, 16]

THE FORMATION OF CLOUDS

Clouds form when the invisible moisture in the air condenses in the form of water droplets or ice crystals. The cloud particles remain in suspension in the air in the form of a colloid. The air currents and turbulent eddies are sufficient to sustain the small cloud droplets. If, however, the droplets grow to such a size and weight that they can no longer be held aloft by the turbulence of the air, they will fall from the cloud as rain or snow. The direct condensation from water vapor to liquid may take place even at temperatures below 32° F. When water vapor changes directly into ice crystals, the process is called sublimation. This will form a cloud of ice crystals. Pure ice crystal clouds are usually formed at temperatures well below 32° F.

There are a number of ways by which the air may be cooled to bring about saturation and cloud formation, and these are the bases of textbooks in weather forecasting and theoretical meteorology.

Cloud observations should be made accurately with respect to the type, the amount, the direction of movement, and the height, as the first step in providing the forecaster with information upon which to base forecasts of cloud growth or dissipation.

CLOUD CLASSIFICATION

A system of classification, based upon the forms and relative altitudes of clouds, has been selected. This system is discussed below. Some of the physical processes producing clouds are also considered. Classification of clouds under a uniform system

22

throughout the world was one of the most progressive steps in modern meteorology.

In 1929 the International Meteorological Organization met in Copenhagen, Denmark, and adopted the present cloud code and definitions of cloud forms. The organization published their results in the *International Atlas of Clouds and States of the Sky*, of which the third edition was published in Paris, France, in 1932.

Family	Mean Lower Level *	Mean Upper Level *	Genus	Form	Abbreviation
A. High clouds	6,000 m. 20,000 ft.	Tropopause	1. Cirrus	Sheets of filaments or scales	CI CC
			2. Cirrocumulus		
			3. Cirrostratus	Continuous sheet	CS
B. Middle clouds	2,000 m. 6,500 ft.	6,000 m. 20,000 ft.	4. Altocumulus	Heap cloud filament sheets	AC
			5. Altostratus	Continuous sheet	AS
C. Low clouds	Near ground	2,000 m. 6,500 ft.	6. Stratocumulus	Heap cloud filament sheet	SC
			7. Stratus	Continuous sheet	ST
			8. Nimbostratus		NS
D. Clouds with vertical development	500 m. 1,600 ft.	6,000 m. 20,000 ft.	9. Cumulus	Heap clouds	CU
			10. Cumulonimbus		CB

* Above ground.

The United States Weather Bureau in 1938 adopted the international system of classification and issued a circular to their observers entitled *Codes for Cloud Forms and States of the Sky*. This circular was intended for observers who report clouds in synoptic telegraphic messages in the International Numeral Weather Code.[9, 16]

Clouds could also be classified according to the method of formation. Clouds formed by cooling due to gradual upslope motion and expansion of air along fronts might be classed as frontal clouds. Clouds formed by cooling due to expansion of air in strong, rising currents might be classed as one type of air mass cloud. This method of classification is, however, not suitable for universal observations, as it would require the observer to have a knowledge of the physical processes that were taking place in the atmosphere.

Clouds may appear in the following forms: (*a*) isolated, heap clouds with vertical development during their formation and a spreading out when they are dissolving; (*b*) sheet clouds, which are divided into filaments, scales, or rounded masses, and which are stable or in the process of disintegration; (*c*) more or less continuous cloud sheets, often in the process of formation or growth.

The international system classified clouds in four families and ten genera. The table on page 23 gives the *mean* heights above the general ground level of the cloud, applicable to middle latitudes.

HIGH CLOUDS

Cirrus

Cirrus clouds are defined as detached clouds of delicate and fibrous appearance, without shading, generally white, often of a silky appearance.

Cirrus appears in the most varied forms: isolated tufts, lines drawn across a blue sky, branching featherlike plumes, curved lines ending in tufts, etc.; they are often arranged in bands which cross the sky like meridian lines, and which, owing to the effect of perspective, converge to a point on the horizon, or to two opposite points (cirrostratus and cirrocumulus often take part in the formation of these bands).

Cirrus clouds are always composed of ice crystals, and their transparent character depends upon the degree of separation of the crystals.

As a rule when these clouds cross the sun's disk they hardly diminish its brightness. But when they are exceptionally thick they may veil its light and obliterate its contour. This would also be true of patches of altostratus, but cirrus is distinguished by the dazzling and silky whiteness of its edges.

Halos are rather rare in cirrus.

Sometimes isolated wisps of snow seen against the blue sky resemble cirrus; they are of a less pure white and less silky than cirrus; wisps of rain are definitely gray, and a rainbow, should one be visible, shows their nature at once, for a rainbow cannot be produced in cirrus.

Before sunrise and after sunset, cirrus is often colored bright yellow or red. These clouds are lit up long before other clouds and fade out much later; some time after sunset they become gray. At all hours of the day cirrus near the horizon is often yellowish; this color is due to distance and to the great thickness of air traversed by the rays of light.

Cirrus, being in general more or less inclined to the horizontal, tends less than other clouds to become parallel to the horizon, under the effect of perspective, as the horizon is approached; often, on the contrary, it seems to converge to a point on the horizon.

Among the principal species one may note:

CIRRUS FILOSUS. More or less straight or irregularly curved filaments (neither tufts nor little hooks and without any of the parts being fused together).

CIRRUS UNCINUS. Cirrus in the shape of a comma; the upper part either ends in a little tuft or is pointed.

CIRRUS DENSUS. Cirrus clouds with such thickness that without care an observer might mistake them for middle or low clouds.

CIRRUS NOTHUS. Cirrus proceeding from a cumulonimbus and composed of the debris of the upper frozen parts of these clouds.

Ordinary cirrus may appear in many very different varieties. One may particularly note the forms floccus and vertebratus which are really aspects of the varieties cumuliformis and undulatus radiatus, respectively.

Cirrocumulus

Cirrocumulus clouds are defined as a cirriform layer or patch composed of small white flakes or of very small globular masses, usually without shadows, which are arranged in groups or lines, or more often in ripples resembling those of the sand on the seashore.

In general, cirrocumulus represents a degraded state of cirrus and cirrostratus, both of which may change into it. In this event the changing patches often retain some fibrous structure in places.

Real cirrocumulus is uncommon. It must not be confused with small altocumulus on the edges of altocumulus sheets. There are, in fact, all states of transition between cirrocumulus and altocumulus proper, as is only to be expected since the process of formation is the same. In the absence of any other criterion the term cirrocumulus should only be used when:

1. There is evident connection with cirrus or cirrostratus.
2. The cloud observed results from a change in cirrus or cirrostratus.
3. The cloud observed shows some of the characteristics of ice crystal clouds which will be found enumerated under cirrus.

Clear rifts are often seen in a sheet of cirrocumulus.

Cirrostratus

Cirrostratus clouds are defined as a thin whitish veil which does not blur the outlines of the sun or moon but usually gives rise to halos. Sometimes it is quite diffuse and merely gives the sky a milky look; sometimes it more or less distinctly shows a fibrous structure with disordered filaments.

A sheet of cirrostratus which is very extensive, though in places it may be interrupted by rifts, nearly always ends by covering the whole sky. The border of the sheet may be straight edged and clear cut, but more often it is ragged or cut up.

During the day, when the sun is sufficiently high above the horizon, the sheet is never thick enough to prevent shadows of objects on the ground.

A milky veil of fog (or thin stratus) is distinguished from a veil of cirrostratus of a similar appearance by the halo phenomena which the sun or the moon nearly always produces in a layer of cirrostratus.

The following are the principal halo phenomena: A circle of 22° radius around the sun or moon—roughly, the angle subtended by the hand placed at right angles to the arm when the arm is extended; this halo is sometimes, but rarely, accompanied by one of 46° radius. Parhelia, paraselenae (mock suns or mock moons), luminous patches, often showing prismatic colors, a little over 22° from the sun or moon and at the same elevation. A luminous column, e.g., sun pillar, extending vertically above and below the luminary.

Often only small fragments of these appearances are visible but they are none the less characteristic of high clouds.

What has been said above of the transparent character and colors of cirrus is true to a great extent of cirrostratus.

Cirrostratus has two principal aspects which correspond to the two following species:

CIRROSTRATUS NEBULOSUS. A uniform nebulous veil, sometimes very thin and hardly visible, sometimes relatively dense, but always without definite details and usually with halo phenomena.

CIRROSTRATUS FILOSUS. A white fibrous veil, where the strands are more or less definite, often resembling a sheet of cirrus densus from which indeed it may originate.

MIDDLE CLOUDS

Altocumulus

Altocumulus clouds are defined as a layer (or patches) composed of laminae or rather flattened globular masses, the smallest elements of the regularly arranged layer being fairly small and thin, with or without shading. These elements are arranged in groups, in lines, or waves, following one or two directions, and are sometimes so close together that their edges join.

The thin and translucent edges of the elements often show irisations which are rather characteristic of this class of cloud.

From the definition it follows that altocumulus comprises the subgenera:

ALTOCUMULUS TRANSLUCIDUS. Altocumulus formed of elements whose color—from dazzling white to dark gray—and whose thickness vary much from one example to another, or even in the same layer; the elements are more or less regularly arranged and distinct. In the definition of the elements it is the variation in the transparency of the layer, variable from one point to another, that plays the essential part. There appears in the interstices either the blue of the sky, or at least a marked lightening of the layer of cloud due to a thinning out.

ALTOCUMULUS OPACUS. An altocumulus sheet which is continuous, at least over the greater part of the layer, and consisting of dark and more or less irregular elements, in the definition of which transparency does not play a great part, owing to the thickness and density of the layer; but the elements show a real relief on the lower surface of the cloud sheet.

The limits within which altocumulus is met are very wide.

At the greatest heights, altocumulus made up of small elements resembles cirrocumulus; altocumulus, however, is distinguished

by not possessing any of the following characters of cirrocumulus:
1. Connection with cirrus or cirrostratus.
2. An evolution from cirrus or cirrostratus.
3. Properties due to physical structure (ice crystals) enumerated under cirrus.

At lower levels, where altocumulus may be derived from a spreading out of the tops of cumulus clouds, it may easily be mistaken for stratocumulus; the convention is that the cloud is altocumulus if the smallest, well-defined, and regularly arranged elements which are observed in the layer (leaving out the detached elements generally seen on the edges) are not greater than 10 solar diameters in their smallest diameters, i.e., approximately the width of three fingers when the arm is held extended.

When the edge or a thin semitransparent patch of altocumulus passes in front of the sun or moon a corona appears close up to (within a few degrees of) them. This corona is a colored ring with red outside and blue inside; the colors may be repeated more than once. This phenomenon is infrequent with cirrocumulus, and only the higher forms of stratocumulus show it.

Irisation, mentioned above, is a phenomenon of the same type as the corona; it is characteristic of altocumulus as distinguished from cirrocumulus or stratocumulus.

Altocumulus clouds often appear at different levels at one and the same time. Often, too, they are associated with other types of cloud.

The atmosphere is often hazy just below altocumulus clouds.

When the elements of a sheet of altocumulus fuse together and make a continuous layer altostratus or nimbostratus is the result. On the other hand a sheet of altostratus can change into altocumulus. It may happen that these two aspects of a cloud sheet may alternate with each other during the whole course of a day. It is also not rare to have a layer of altocumulus coexisting with a veil resembling altostratus at a height very little less than the altocumulus.

It is interesting to note that one may often observe filiform descending trails to which the name virga has been given.

Among the principal species one may note:

ALTOCUMULUS CUMULOGENITUS. This is an altocumulus cloud formed by the spreading out of the tops of cumulus, the lower parts of the cumulus clouds having melted away; the layer in the first stages of its growth has the appearance of altocumulus opacus.

An important variety of altocumulus should be noted, namely, altocumulus cumuliformis, which has two different aspects:

ALTOCUMULUS FLOCCUS. Tufts resembling small cumulus clouds without a base and more or less ragged.

ALTOCUMULUS CASTELLATUS. Cumuliform masses with more or less vertical development, arranged in a line, and resting on a common horizontal base, which gives the cloud a crenelated appearance.

The caps or hoods which form above a cumulus by the uplift of a damp layer, and which may be pierced by the tops of the cumulus, are considered a detail of cumulus and denoted by the term pileus attached to the name cumulus; but in reality they are aberrant forms of altocumulus translucidus. Moreover, the formation of similar clouds, independent of cumulus, can be effected by a current of air rising over a mountain or any obstacle. They are then named altocumulus, and they are classed, on account of their form, with the variety lenticularis.

Altostratus

Altostratus clouds are defined as a striated or fibrous veil, more or less gray or bluish. This cloud is like thick cirrostratus but without halo phenomena; the sun or moon shows vaguely, with a faint gleam, as though through ground glass. Sometimes the sheet is thin, with forms intermediate with cirrostratus. Sometimes it is very thick and dark, sometimes even completely hiding the sun or moon. In this event differences of thickness may cause relatively light patches between very dark parts; but the surface never shows real relief, and the striated or fibrous structure is always seen in places in the body of the cloud.

Every form is observed between high altostratus and cirrostratus on the one hand, and low altostratus and nimbostratus on the other.

Rain or snow may fall from altostratus (altostratus precipitans), but when the rain is heavy the cloud layer will have grown thicker and lower, becoming nimbostratus; but heavy snow may fall from a layer that is definitely altostratus.

From the definition of altostratus it follows that there are three subgenera:

ALTOSTRATUS TRANSLUCIDUS. A sheet of altostratus resembling thick cirrostratus; the sun and the moon show as through ground glass.

ALTOSTRATUS OPACUS. An opaque layer of altostratus of variable thickness which may entirely hide the sun, at any rate, in parts, but shows a fibrous structure in some parts.

ALTOSTRATUS PRECIPITANS. A layer of opaque altostratus which has not yet lost its fibrous character, and from which there are light falls of rain or snow, either continuous or intermittent. This precipitation may not reach the ground, in which case it forms virgae.

The limits between which altostratus may be met with are fairly wide (about 5,000 to 2,000 meters).

A sheet of high altostratus is distinguished from a rather similar sheet of cirrostratus by the circumstance that halo phenomena are not seen in altostratus, nor are the shadows of objects on the ground visible.

A sheet of low altostratus may be distinguished from a somewhat similar sheet of nimbostratus by the following characteristics. Nimbostratus is of a much darker and more uniform gray, and shows nowhere any whitish gleam or fibrous structure; one cannot definitely see the limit of its undersurface, which has a wet look, due to the rain (or snow), which may not reach the ground.

The convention is that nimbostratus always hides the sun and moon in every part of it, whereas altostratus hides them only in places behind its darker portions, but they reappear through the lighter parts.

Careful observation may often detect virgae hanging from altostratus, and these may even reach the ground, causing slight precipitation. If the sheet still has the character of altostratus it will then be called altostratus precipitans, but if not it has become nimbostratus.

A sheet of altostratus, even if it has rifts in places, has a general fibrous (or ground-glass) character. A cloud layer, even a continuous one, which has no fibrous structure, and in which rounded cloud masses may be seen, is classed as altocumulus or stratocumulus according to circumstances.

Altostratus may result from a transformation of a sheet of altocumulus, and on the other hand altostratus may often break up into altocumulus.

Low Clouds

Stratocumulus

Stratocumulus clouds are defined as a layer (or patches) composed of laminae, globular masses or rolls; the smallest of the regularly arranged elements are fairly large; they are soft and gray, with darker parts.

These elements are arranged in groups, in lines, or in waves, alined in one or in two directions. Very often the rolls are so close that their edges join; when they cover the whole sky they have a wavy appearance.

From the definition it follows that stratocumulus comprises two kinds:

STRATOCUMULUS TRANSLUCIDUS. A not very thick layer; in the interstices between its elements either the blue sky appears, or at any rate there are much lighter parts of the cloud sheet, which here is thinned out on its upper surface.

STRATOCUMULUS OPACUS. A very thick layer made up of a continuous sheet of large dark rolls or rounded masses; their shape is seen not by a difference in transparency, but they stand out in real relief from the under surface of the cloud layer.

There are transitional forms between stratocumulus and altocumulus on the one hand and between stratocumulus and stratus on the other.

The difference between stratocumulus and altocumulus is discussed under altocumulus.

It should also be noted that the cloud sheet called altocumulus by an observer at a small height might appear as stratocumulus to an observer at a greater height.

It often happens that stratocumulus is not associated with any clouds of the second or third families; but it fairly often coexists with clouds of the fourth family.

The elements of thick stratocumulus (stratocumulus opacus) often tend to fuse together completely, and the layer can, in certain cases, change into nimbostratus. The cloud is called nimbostratus when the cloud elements of stratocumulus have completely disappeared and when, owing to the trails of falling precipitation, the lower surface has no longer a clear-cut boundary.

Stratocumulus can change into stratus, and vice versa. The stratus being lower, the elements appear very large and very soft, so that the structure of regularly arranged globular masses and waves disappears as far as the observer can see. The cloud will be called stratocumulus as long as the structure remains visible.

Among the principal species may be mentioned:

STRATOCUMULUS VESPERALIS. This name is given to flat, elongated clouds which are often seen to form about sunset as the final product of the diurnal changes of cumulus.

STRATOCUMULUS CUMULOGENITUS. Stratocumulus formed by the spreading out of the tops of cumulus clouds, which later have disappeared; the layer in the early stages of its formation looks like stratocumulus opacus.

The cloud called roll cumulus in England and Germany is designated stratocumulus undulatus; its wave system is in one direction only. It must not be confused with flat cumulus clouds ranged in line. Stratocumulus often has a mammatus (festooned) character; that is to say, there is a high relief on the lower surface where pendant rounded masses or corrugations are observed, which at times look as though they would become detached from the cloud. Care must be taken not to confuse this cloud with some kinds of altostratus opacus whose under surface may appear to be slightly corrugated or mammillated; altostratus opacus is distinguished by its fibrous structure.

Stratus

Stratus clouds are defined as a low uniform layer of cloud, resembling fog, but not resting on the ground.

FRACTOSTRATUS. When this very low layer is broken up into irregular shreds it is designated fractostratus.

A veil of true stratus generally gives the sky a hazy appearance which is very characteristic, but which sometimes may cause confusion with nimbostratus. When there is precipitation the difference is manifest; nimbostratus gives continuous rain (sometimes snow), precipitation composed of drops which may be small and sparse, or else large (at least some of them) and close together, whereas stratus only gives a drizzle, that is to say, small drops very close together. Precipitation falling from upper clouds may, however, fall through a layer of stratus under frontal conditions.

When there is no precipitation a dark and uniform layer of stratus can easily be mistaken for nimbostratus. The lower surface of nimbostratus, however, always has a wet appearance (widespread trailing precipitation, "virga"); it is quite uniform, and definite details cannot be made out; stratus, on the other hand, has a drier appearance, and however uniform it may be it shows some contrasts and some lighter transparent parts, that is, places less dark where the cloud is thinner, corresponding to the interstices between the rolls and globular masses of stratocumulus, but considerably larger, while nimbostratus seems only to be feebly illuminated, as though lit up from within.

Stratus is often a local cloud, and when it breaks up the blue sky is seen.

Fractostratus sometimes originates from the breaking up of a layer of stratus; sometimes it forms independently and develops until it forms a layer below altostratus or nimbostratus, the nimbostratus sometimes being visible through the interstices.

A layer of fractostratus may be distinguished from nimbostratus by its darker appearance and by being broken up into cloud elements. If these elements have a cumuliform appearance in places the cloud layer is called fractocumulus and not fractostratus.

Nimbostratus

Nimbostratus clouds are defined as a low, amorphous, and rainy layer, of a dark gray color, usually nearly uniform; feebly illuminated, seemingly from inside. When it gives precipitation it is in the form of continuous rain or snow.

But precipitation alone is not a sufficient criterion to distinguish the cloud which should be called nimbostratus even when no rain or snow falls from it.

There is often precipitation which does not reach the ground; then the base of the cloud is usually diffuse and looks wet on account of the general trailing precipitation, virga, so that it is not possible to determine the limit of its lower surface.

The usual evolution is as follows: A layer of altostratus grows thicker and lower until it becomes a layer of nimbostratus. Beneath this layer there is generally a progressive development of very low ragged clouds, isolated at first, then fusing together into an almost continuous layer, through the interstices of which the nimbostratus can generally be seen. These very low clouds are called fractocumulus or fractostratus according as they appear cumuliform or stratiform.

Generally the rain falls only after the formation of these very low clouds, which are then hidden by the precipitation or may even melt away under its action. The vertical visibility then becomes very bad.

Under certain conditions the precipitation may precede the formation of fractocumulus or fractostratus, or it may happen that these clouds do not form at all.

Rather rarely a sheet of nimbostratus may form by an evolution from a stratocumulus.

Clouds with Vertical Development

Cumulus

Cumulus clouds are defined as dense clouds with vertical development; the upper surface is dome shaped and exhibits rounded protuberances, while the base is nearly horizontal.

When the cloud is opposite the sun the surfaces normal to the observer are brighter than the edges of the protuberances. When the light comes from the side, the clouds exhibit strong contrasts of light and shade; against the sun, on the other hand, they look dark with a bright edge.

Fractocumulus. True cumulus is definitely limited above and below; its surface often appears hard and clear cut. But one may also observe a cloud resembling ragged cumulus in which the different parts show constant change. This cloud is designated fractocumulus.

Typical cumulus, over land areas, develops on days of clear skies and is due to the currents of diurnal convection; it appears in the morning, grows, and then more or less dissolves again toward the evening.

Cumulus, whose base is generally of a gray color, has a uniform structure; that is to say, it is composed of rounded parts right up to its summit, with no fibrous structure. Even when highly developed, cumulus can produce only light precipitation.

Cumulus, when it reaches the altocumulus level, is sometimes capped with a light, diffuse, and white veil of more or less lenticular shape, with a delicate striated or flaky structure on its edges; it is generally shaped like a bow which may cover several domes of the cumulus and finally be pierced by them. This cloud, which does not constitute a species, is given the name pileus, a cap or hood.

The clouds which form below altostratus or nimbostratus and which can develop into a complete layer, through whose interstices the altostratus or nimbostratus is generally seen, are usually fractostratus; but if they have a cumuliform appearance they should be classed as fractocumulus. They rarely have this appearance during or soon after rain; on the other hand it is frequent at the beginning of the formation of the low cloud and when it breaks up.

Among the principal species one may note:

CUMULUS HUMILIS. Cumulus with little vertical development, and seemingly flattened. These clouds are generally seen in fine weather.

CUMULUS CONGESTUS. Very distended and sprouting cumulus, whose domes have a cauliflower appearance.

Cumulonimbus

Cumulonimbus clouds are defined as heavy masses of cloud, with great vertical development, whose cumuliform summits rise in the form of mountains or towers, the upper parts having a fibrous texture and often spreading out in the shape of an anvil.

The base resembles nimbostratus, and one generally notices virga. This base has often a layer of very low ragged clouds below it (fractostratus, fractocumulus).

Cumulonimbus clouds generally produce showers of rain or snow and sometimes of hail, and often thunderstorms as well.

If the whole of the cloud cannot be seen the fall of a real shower is enough to characterize the cloud as a cumulonimbus.

Even if a cumulonimbus were not distinguished by its shape from a strongly developed cumulus its essential character is evident in the difference of structure of its upper parts, when they are visible (fibrous structure and cumuliform structure). Masses of cumulus, however heavy they may be, and however great their vertical development, should never be classed as cumulonimbus unless the whole or a part of their tops is transformed or is in process of transformation into a cirrus mass.

Although the upper cirriform parts of a cumulonimbus may take on very varied shapes, sometimes they spread out into the form of an anvil. To this interesting feature the name incus is given.

In certain types of cumulonimbus, which are especially common in spring in moderately high latitudes, the fibrous structure extends to nearly the whole cloud mass, so that the cumuliform parts almost wholly disappear; the cloud is reduced to a mass cf cirrus and of virga.

The veil cloud pileus is seen with cumulonimbus clouds as with cumulus.

When a cumulonimbus covers nearly all the sky the base alone is visible and resembles nimbostratus, with or without fracto-stratus or fractocumulus below. The difference between the base of a cumulonimbus and a nimbostratus is often rather difficult to make out. If the cloud mass does not cover all the sky, and if even small portions of the upper parts of the cumulonimbus appear, the difference is evident. If not it can be made out only if the preceding evolution of the clouds has been followed, or if precipitation occurs; the precipitation of a cumulonimbus is violent and intermittent (showers) as opposed to the relatively gentle and continuous precipitation of a nimbostratus.

The front of a thunder cloud of great extent is sometimes accompanied by a roll cloud of a dark color in the shape of an arch, of a frayed-out appearance, and circumscribing a part of the sky of a lighter gray. This cloud is named arcus and is nothing more or less than a particular example of fractocumulus or fractostratus.

Fairly often a mammatus structure appears in cumulonimbus, either at the base, or on the lower surface of the lateral parts of the anvil.

When a layer of menacing cloud covers the sky and both virga and mammatus structure are seen it is a sure sign that the cloud is the base of a cumulonimbus, even in the absence of all other signs.

Cumulonimbus is a real factory of clouds; it is responsible in great measure for the clouds in the rear of disturbances. By the spreading out of the high parts and the melting away of the underlying parts, cumulonimbus can produce thick sheets of altocumulus or stratocumulus (spreading out of the cumuliform parts) and dense cirrus (spreading out of the cirriform parts).

Among the principal species may be noted:

CUMULONIMBUS CALVUS. Cumulonimbus characterized by the thunderstorm or the shower that it causes, or by virga, but in which no cirriform parts can be made out. Nevertheless the freezing of the upper parts has already begun; the tops are beginning to lose their cumulus structure, that is, their rounded outlines and clear-cut contours; the hard and "cauliflower" swellings soon become confused and melt away so that nothing can be seen in the white mass but more or less vertical fibers. The freezing, accompanied by the change into a fibrous structure, often goes on very rapidly.

CUMULONIMBUS CAPILLATUS. Cumulonimbus which displays distinct cirriform parts, having sometimes, but not always, the shape of an anvil.

PRINCIPAL VARIETIES OF CLOUDS

The chief varieties common to different genera are as follows:

Fumulus

At all levels, from cirrus to stratus, a very thin veil may form, so delicate that it may be almost invisible. These veils seem to be most frequent on hot days and in low latitudes. Occasionally they may be observed to thicken rapidly, forming clouds easily visible, especially cirrus and cumulus. The clouds thus produced seem unstable, however, and usually melt away soon after their formation.

Cirrus fumulus must not be confused with cirrostratus nebulosus, which is much more stable and does not show the phenomenon of the formation and subsequent rapid disappearance of cirrus clouds.

Lenticularis

Clouds of an ovoid shape, with clean-cut edges, and sometimes irisations, especially common on days of strong, dry winds in rough country. This form exists at all levels from cirrostratus to stratus.

Cumuliformis

The rounded form resembling cumulus which the upper parts of other clouds may sometimes assume. This may be seen at all levels from cirrus to stratus.

Mammatus

This description is given to all clouds whose lower surfaces form pouches. This form is found especially in stratocumulus and in cumulonimbus, either at the base, or even more often on the lower surface of anvil projections. It is also found, though rarely, in cirrus clouds, probably when they have originated in the anvil of a dispersing cumulonimbus.

Undulatus

This term is applied to clouds composed of elongated and parallel elements, like waves of the sea. There is sometimes an appearance of two distinct systems, as when the cloud is divided into rounded masses by undulations in two directions.

Radiatus

This term is applied to clouds in parallel bands (polar bands), which owing to perspective seem to converge to a point on the horizon, or to two opposite points if the bands cross the whole sky. The point is called the radiant or vanishing point.

Chief Casual Details

The chief casual varieties are the following:

VIRGA. Wisps or falling trails of precipitation; applied principally to altocumulus and altostratus.

U.S.W.B. *photograph by H. T. Floreen.*

FIG. 7.   Cumulus of fine weather.

PILEUS. A cap or hood; applied principally to cumulus or cumulonimbus.

INCUS. Anvil; upper part of cumulonimbus.

ARCUS. Arch cloud; usually associated with cumulonimbus.

Photograph by C. E. Deppermann at Manila.

Fig. 8.   Heavy and swelling cumulus without anvil top.

U.S.W.B. photograph by P. A. Miller at Washington, D. C.

Fig. 9.   Cumulonimbus.

Clouds

U.S.W.B. photograph.

FIG. 10.   Stratocumulus formed by spreading out of cumulus.

U.S.W.B. photograph.

FIG. 11.   Layer of stratocumulus.

U.S.W.B. photograph.

Fig. 12.   Layer of stratus.

U.S.W.B. photograph.

Fig. 13.   Low, ragged, dark gray clouds of bad weather.

FIG. 14.   Aerial view of heavy and swelling cumulus.

FIG. 15.   Heavy cumulus or cumulonimbus and stratocumulus.

Fig. 16. Cumulonimbus with underlying low, ragged clouds of bad weather.

Fig. 17. Typical thin altostratus.

FIG. 18.   Typical thick altostratus.

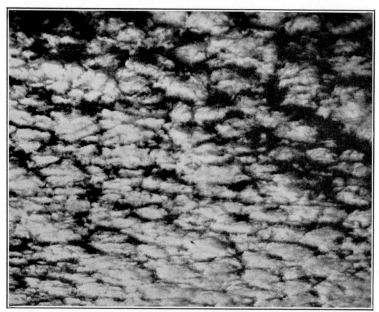

FIG. 19.   Sheet of altocumulus at one level.

U.S.W.B. photograph.

Fig. 20.   Altocumulus in small isolated patches more or less lenticular in shape.

U.S.W.B. photograph by Willard S. Wood.

Fig. 21.   Altocumulus arranged in parallel bands spreading over the sky.

U.S.W.B. photograph.

Fig. 22. Altocumulus formed by spreading out of the tops of cumulus.

U.S.W.B. photograph.

FIG. 23. Altocumulus associated with altostratus.

U.S.W.B. photograph by L. E. Johnson.

FIG. 24. Altocumulus castellatus.

FIG. 25.   Altocumulus at different levels.

FIG. 26.   Delicate cirrus not forming a continuous layer.

FIG. 27.   Dense cirrus derived from an anvil.

FIG. 28.   Delicate cirrus in the form of hooks ending in a little claw or up-
turned end.

U.S.W.B. photograph by A. J. Weed.

FIG. 29.　Cirrus and cirrostratus not higher than 45° above the horizon.

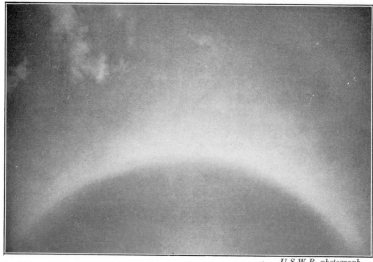

U.S.W.B. photograph.

FIG. 30.　Cirrostratus showing a 22° halo.

U.S.W.B. photograph by Louise A. Boyd.

Fig. 31. Cirrostratus covering all the sky.

Fig. 32. Cirrostratus not increasing in amount, and not covering all the sky.

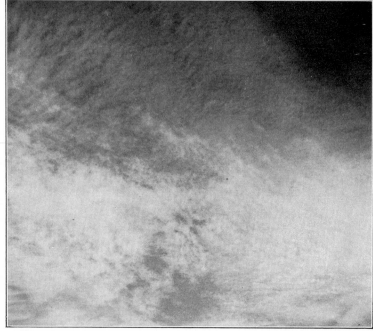

U.S.W.B. photograph.

Fɪɢ. 33.   Cirrocumulus predominating associated with the cirrus mass.

Aspect of the Sky

Before considering methods of determining the proportion of the sky covered by clouds, it is necessary to discuss something of the apparent shape of the sky.[2]   The shape of the sky has been

Fɪɢ. 34.   Perspective effects with scattered cumulus clouds (after Humphreys).

described as a great dome; when covered by high clouds, it appears to be low and flat.   To the observer, the clouds directly overhead at the zenith position may spread out in all directions to vanish from sight on the circular rim of the horizon. Narrow bands of clouds in a blue sky may stretch from one horizon up over the dome and down to the other horizon. Although the bands of clouds may be parallel, they will appear to converge at the horizons. This is due to perspective, which is easily illustrated by the converging rails as one looks down a long, straight railroad track.

Another effect due to perspective, when scattered cumulus clouds are observed against the sky, is the appearance of complete cloud cover at a distance and scattered clouds with blue sky areas nearly overhead. The more distant clouds may also appear to be lower, although the bases of all are at the same height.

AMOUNT OF CLOUDS (SKY COVER) [10]

Definition

The proportion of the sky covered by clouds is based upon a scale from 0 to 10 and is given in tenths. A sky completely filled with clouds, with no blue sky visible, is said to have $10/10$'s cloud cover. With $10/10$'s clouds, the sky is *overcast*. The United States Weather Bureau has adopted the following terminology for use in airway weather reports:

Clear   When there are no clouds present, or less than $1/10$ of the sky is covered by clouds.

Scattered   When from $1/10$ to $5/10$'s, inclusive, of the sky is covered by clouds.

Broken   When more than $5/10$'s of the sky but not more than $9/10$'s of the sky is covered by clouds.

Overcast   When more than $9/10$'s of the sky is covered by clouds.

The total amount of cloud cover is obtained by viewing the sky and projecting the clouds against the dome of the sky. This projected amount of clouds compared to blue sky visible, or, at night, star-filled clear areas, represents the sky cover. The sum of the number of tenths of cloud cover plus the sum of the number of tenths of open sky should always equal $10/10$'s.

Method of Estimating Sky Cover

In estimating the sky cover, the sky dome may be mentally divided into quadrants by means of diameters at right angles to each other. The number of tenths of each quadrant may be estimated separately, and the mean of the four values may then be taken as the total sky cover. Selection of quadrants may be based upon the distribution of clouds in the sky so that each area may be easily estimated. It is recognized that broken cumulus clouds will give an appearance of running together at a distance because of viewing the sides as well as the bases. The rule of projection against the dome of the sky should, however, still be used.

When a number of layers of clouds are present, each layer should be projected against the sky separately. Frequently the upper layer must be observed through breaks in a lower layer. Under these conditions it is necessary to spend sufficient time watching the upper layer through the moving clear spaces between the lower clouds to determine the number of tenths covered. It is possible to determine that a higher overcast exists above a layer that covers as much as $\frac{6}{10}$'s or $\frac{7}{10}$'s of the sky. For example, we may observe $\frac{10}{10}$'s AS above $\frac{6}{10}$'s CU.

If the sky is being watched carefully, an observer may be able to see a lower deck of clouds move in or actually form below an upper deck. During and immediately following frontal passages, clouds may appear in many levels, and the whole appearance of the sky is chaotic. If in this event the cloud masses do not cover sufficient area in themselves to be accurately estimated, they may be grouped together to determine the total sky cover and may be described as numerous layers.

Observation of cloud layers may be simplified by studying the cloud motions, as it is rare to have clouds at all levels moving at the same speed and in the same direction.

Cloud Height [10]

Definition

Cloud height is the distance from the ground to the base of the cloud. The United States Weather Bureau has used the term *ceiling* to define a value which involves a number of layers of clouds. It may be defined as follows: Ceiling is the height above ground of the lowest clouds reported as broken or overcast. In other words, ceiling is the lowest height above the ground at which all cloud layers at and below that level cover more than half of the sky. The Weather Bureau defines *unlimited ceiling* as: (1) no broken or overcast layer is reported; or (2) the base of the lowest reported broken or overcast layer is higher than 9,750 feet above ground. *Ceiling zero* is reported when the ceiling is 50 feet or less.

Methods of Measuring Cloud Height

Present-day methods for measuring cloud height are: ceiling balloon, ceiling light projector, clinometer, and the ceilometer. Other methods for obtaining the height of clouds are by airplane,

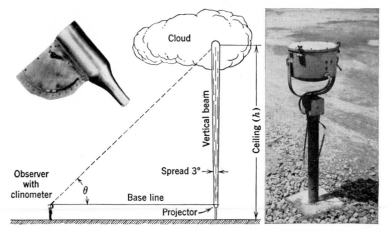

Fig. 35. Ceiling light and clinometer determination of cloud height.

triangulation, range finders, pilot balloons, intersections with hills or towers, and estimating by application of the dew-point formula.

CEILING LIGHT PROJECTOR AND CLINOMETER. Making ceiling measurements by the ceiling light method requires a point for observation, usually near the observatory, and a searchlight located 500 feet or 1,000 feet from the observation point. The projector is usually north of the observer (in the northern hemisphere) to avoid interference by moonlight.

The projector is a small searchlight which projects a narrow beam of light of less than 3° spread onto the base of the cloud. It may be pointed vertically or at an angle to the vertical toward the observer. The vertical method is used in the United States. The angular elevation of the light spot on the cloud base is measured by a clinometer or an alidade. The clinometer is a sighting tube with cross wires placed at the outer end. A quadrant scale graduated in degrees from 0 to 90 is mounted on the side. A weighted pendant, which hangs vertically when free, slides along the arc of the quadrant scale. When the clinometer is pointed up to the spot the observer may clamp the pendant to the scale by means of a clutch head screw. The angle above the horizontal may then be read on the scale. The method of computation is by simple triangle as shown in Fig. 35 or

$$\text{Height} = \text{Base line} \times \tan \theta$$

If the base of the cloud is uneven or if it gradually thins out near the base, the spot may not be clearly reflected. With thin clouds there may be considerable penetration by the light beam before it is lost. The point of first entry, or low spot, is usually taken as the cloud base. When the base of the cloud is ragged, hanging cloud masses may interfere with the line of sight and it may be necessary to watch the cloud until a clearly defined spot is observed. Also, it may be desirable to average the high and low angles to obtain a representative figure for the cloud base height. Under these conditions further description of the character of the cloud base may be given.

The accuracy of measurement by ceiling light projector methods depends upon the length of base line, the accuracy of reading the clinometer, and the evenness of the base of the cloud. Using a 500-foot base line and with a clinometer reading which is uncertain to 2° for a cloud height of 1,800 feet, the error would be 230 feet; for a 1,500-foot base line the error would be 130 feet.

With projectors used by the United States Weather Bureau cloud heights may be obtained under ideal conditions up to 15,000 feet. The errors of measurement become large, however, at the higher elevations. With a 1,000-foot base line an error of 1° in reading the clinometer will give an error of about 2,000 feet at 10,000 feet, and an error of over 3,000 feet at 15,000 feet elevation.

THE CEILOMETER. A new device now being installed at Weather Bureau airport stations for cloud-height measurement is the ceilometer. The principle is identical to that of the ceiling light projector and clinometer except that a light modulated to a known frequency is projected onto the base of the cloud. Instead of the clinometer, a receiver uses a photoelectric tube as the seeing eye.

The photoelectric element is mounted in a reflector behind the diaphragm at the focal point. The receiver is mounted so that it may be rotated in a vertical plane and the angle from the horizontal read on an arc similar to the clinometer. Because of the modulated light, a selective electronic unit reacts to any light, direct or reflected, which the projector sends up. In this way the ceilometer may be used in the daytime as well as at night. The accuracy of measurements by this unit is of the same order as the ceiling light method. Cloud heights to 10,000 feet may be measured during the daytime and to 20,000 feet at night.

CEILING BALLOONS. Balloons weighing about 10 grams are inflated with hydrogen gas to a free lift of 40 grams or with helium gas to a free lift of 45 grams. When released, these balloons rise at an average rate shown in Table III–1. The balloon is observed, and the time from release to entry into the cloud base is noted. The height of the cloud base is then the time multiplied by the ascensional rate. Cloud heights up to 2,000 or 3,000 feet may be measured in this way. The ascensional rate of ceiling balloons varies considerably, however, and the results obtained from a single measurement are not too reliable. Table III–1 shows the variation under ideal cloud conditions with complete overcast. When the clouds become scattered or even broken, the vertical motions in the lower atmosphere make the ascensional rate of the balloons very erratic, and this method should not be used.

The equipment necessary for this work consists of: (1) compressed hydrogen or helium gas in cylinders, usually containing 200 cubic feet; (2) hydrogen or helium regulator; (3) Brady free-lift device; (4) timepiece, indicating seconds.

The balloons are usually furnished in two colors, red and purple. In general, the darker color is best seen against a dark background.

A specially constructed inflation room is usually located as conveniently as possible to the place of observation. This room should contain a rack for helium tanks and a shelf for inflating the balloon. The tank is fitted with a shutoff valve and a reducing valve, or regulator, having two gages, one registering the amount of pressure in the tank and the other the pressure of the gas flowing into the balloon. A low-pressure valve on the regulator regulates the flow of gas. A $\frac{1}{4}$-inch rubber tube leads from the regula-

tor to a three-way petcock. A small, ⅛-inch rubber tube leads to
the Brady free-lift device. (See Fig. 36.)

The Brady device is placed on a shelf large enough to permit the
small rubber tubing to rest thereon.

The neck of the balloon is stretched over the inflation nozzle.
All air is expelled by opening the three-way petcock and rolling
the balloon to force the air out through the petcock. The infla-

Fig. 36.   Ceiling balloon and Brady free-lift device.

tion is then begun by slowly turning the regulator handle to the
right until a steady flow is maintained. The balloons should be
inflated slowly until the device is just lifted from the supporting
surface. When removing the balloon from the device, the neck
should be twisted several times and secured by soft cotton twine
or by a circular band cut from the end of the neck of the
balloon.

The inflated balloon should be released from a point free from
obstructions, such as houses or trees. Keep the balloon in sight
during its ascent. The interval of time between the releasing of
the balloon and its disappearance in the clouds should be accurately determined.

Table III–1 shows the average height of oval and spherical ceiling balloons at the end of each half minute of elapsed time.

TABLE III–1

Altitude, feet

| Time, minutes | Oval balloon (45 grams free lift)[1] | Spherical balloons | | |
		75 grams free lift [1]	45 grams free lift [1]	40 grams free lift [2]
½	205	280	250	250
1	410	560	500	480
1½	615	840	730	670
2	820	1,120	960	850
2½	1,005	1,400	1,190	1,030
3	1,190	1,680	1,420	1,210
3½	1,375	1,945	1,650	1,390
4	1,560	2,210	1,880	1,570
4½	1,745	2,475	2,090	1,750
5	1,930	2,740	2,300	1,930
5½	2,115	3,005	2,510	2,110
6	2,300	3,270	2,720	2,290
6½	2,485	3,545	2,930	2,470
7	2,670	3,820	3,140	2,650
7½	2,855	4,095	3,350	2,830
8	3,040	4,370	3,560	3,010
8½	3,225	4,645	3,770	3,190
9	3,410	4,920	3,980	3,370
9½	3,595	5,195	4,190	3,550
10	3,780	5,470	4,400	3,730

[1] Inflated with helium. [2] Inflated with hydrogen.

There are many factors, such as the shape of balloon, free lift, turbulence, and cloudiness, which affect the ascensional rates of ceiling balloons. Consequently, ceiling heights obtained by such means cannot be considered "measured" heights. The observer must use judgment in accepting the height indicated by balloons as the true cloud height. From a number of double theodolite observations on ceiling balloons, it has been established that, if a large number of balloons were released, the average ascensional rate would closely approximate the averages given in Table III–1. However, any individual balloon can depart from the average values by a considerable amount, as shown in Tables III–3

TABLE III-2

Ceiling-Height Table for the Two Types of Ceiling Balloons in General Use,
Based on Average Ascensional Rates as Shown in Table III-1

Range in Time

Report-able Height, feet	Oval balloon (45 grams free lift)		Spherical balloon (75 grams free lift)	
	From	To	From	To
0	0′ 0″	0′ 7″	0′ 0″	0′ 5″
100	0′ 8″	0′ 21″	0′ 6″	0′ 15″
200	0′ 22″	0′ 36″	0′ 16″	0′ 26″
300	0′ 37″	0′ 51″	0′ 27″	0′ 37″
400	0′ 52″	1′ 6″	0′ 38″	0′ 48″
500	1′ 7″	1′ 20″	0′ 49″	0′ 58″
600	1′ 21″	1′ 35″	0′ 59″	1′ 9″
700	1′ 36″	1′ 49″	1′ 10″	1′ 20″
800	1′ 50″	2′ 5″	1′ 21″	1′ 31″
900	2′ 6″	2′ 21″	1′ 32″	1′ 41″
1,000	2′ 22″	2′ 37″	1′ 42″	1′ 52″
1,100	2′ 38″	2′ 53″	1′ 53″	2′ 3″
1,200	2′ 54″	3′ 10″	2′ 4″	2′ 14″
1,300	3′ 11″	3′ 26″	2′ 15″	2′ 24″
1,400	3′ 27″	3′ 42″	2′ 25″	2′ 35″
1,500	3′ 43″	3′ 58″	2′ 36″	2′ 46″
1,600	3′ 59″	4′ 15″	2′ 47″	2′ 57″
1,700	4′ 16″	4′ 31″	2′ 58″	3′ 8″
1,800	4′ 32″	4′ 47″	3′ 9″	3′ 19″
1,900	4′ 48″	5′ 3″	3′ 20″	3′ 30″
2,000	5′ 4″	5′ 19″	3′ 31″	3′ 42″
2,100	5′ 20″	5′ 35″	3′ 43″	3′ 53″
2,200	5′ 36″	5′ 52″	3′ 54″	4′ 4″
2,300	5′ 53″	6′ 8″	4′ 5″	4′ 15″
2,400	6′ 9″	6′ 24″	4′ 16″	4′ 27″
2,500	6′ 25″	6′ 40″	4′ 28″	4′ 38″
2,600	6′ 41″	6′ 57″	4′ 39″	4′ 49″
2,700	6′ 58″	7′ 13″	4′ 50″	5′ 0″
2,800	7′ 14″	7′ 29″	5′ 1″	5′ 12″
2,900	7′ 30″	7′ 45″	5′ 13″	5′ 23″
3,000	7′ 46″	8′ 1″	5′ 24″	5′ 34″
3,100	8′ 2″	8′ 17″	5′ 35″	5′ 46″
3,200	8′ 18″	8′ 34″	5′ 47″	5′ 57″
3,300	8′ 35″	8′ 50″	5′ 58″	6′ 8″
3,400	8′ 51″	9′ 6″	6′ 9″	6′ 19″
3,500	9′ 7″	9′ 22″	6′ 20″	6′ 30″
3,600	9′ 23″	9′ 39″	6′ 31″	6′ 41″
3,700	9′ 40″	9′ 55″	6′ 42″	6′ 52″
3,800	9′ 56″	10′ 11″	6′ 53″	7′ 3″
3,900	10′ 12″	10′ 27″	7′ 4″	7′ 14″
4,000	10′ 28″	10′ 43″	7′ 15″	7′ 25″

and III–4. (For notes on estimating cloud height see Chapter XIII.)

TABLE III–3

Percentage of Observations within Specified Departures from Average Altitudes Using Oval Balloons with 45 Grams Free Lift

Minute	Average Altitude, feet	±100	±200	±300	±400	±500	±600	±700	±800	±900
1	410	77	97	100	..	...	..	...	..	...
2	820	42	78	94	98	100	..	...	..	...
3	1,190	39	69	87	93	98	99	100	..	...
4	1,560	34	69	83	93	96	99	99	99	100

TABLE III–4

Percentage of Observations within Specified Departures from Average Altitudes Using Spherical Balloons with 75 Grams Free Lift

Minute	Average Altitude, feet	±100	±200	±300	±400	±500	±600	±700	±800
1	560	92	100	..	...	...	...	..	...
2	1,120	67	96	98	100	...	...	..	...
3	1,680	58	88	98	100	...	...	..	...
4	2,210	44	88	93	98	100	...	..	...
5	2,740	46	73	89	95	95	100	..	...
6	3,270	47	72	83	89	92	97	97	100

DIRECTION AND SPEED OF MOVEMENT OF CLOUDS

Direction

The direction of cloud movement is the direction from which the cloud appears to be coming and may be given to 8 or 16 points of the compass. In order to determine the direction accurately, an instrument called the *nephoscope* may be used. There are two types of nephoscopes, direct-vision and mirror-vision. The simplest type to construct is the direct-vision Besson's comb nephoscope. It consists of a 9-foot vertical rod with a horizontal cross piece 3½ feet long fixed to its top. Along the cross piece equidistant vertical spikes are fixed. The vertical rod is mounted on a pole in such a manner that it may be rotated by means of two ropes attached to another short cross piece fixed to the rod at the lower end. A disk with directions marked off is also fixed to the rod. The height should be adjusted so that a fixed mark on the

U.S.W.B. *photograph by*
Professor Talman.

FIG. 37.   Besson's comb nephoscope.

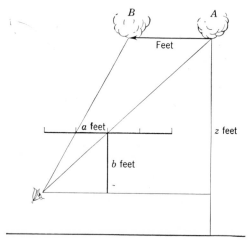

FIG. 38.   Determination of speed of clouds using comb nephoscope.

rod is at the eye level of the observer. To observe the cloud direction, the observer moves about the rod until the cloud to be observed lies in a straight line with the central vertical spike. The rod is then turned by means of the two ropes until the cloud appears to travel along the line of spikes. The direction is then read off on the disk.

Speed of Clouds

In order to determine the speed of the cloud it is necessary to know the height (Z) of the cloud. If a is the distance between the spikes and b is the distance from the upper cross piece to the eye-level mark on the rod, and t is the time it takes for the cloud to move from one spike to the next (Fig. 38), then by similar triangles

$$\frac{A - B}{Z} = \frac{a}{b}$$

If we divide each side by t and multiply by Z we have

$$\frac{A - B}{t} = Z\frac{a}{bt} = \text{Velocity}$$

The left-hand side is the distance in feet traveled by the cloud in t seconds or velocity in feet per second. To convert to miles per hour,

$$1 \text{ foot per second} = \tfrac{30}{44} \text{ miles per hour}$$

For example, 88 feet per second can be converted to miles per hour by dividing by 44 and multiplying by 30, thus

$$\tfrac{88}{44} \times 30 = 60 \text{ miles per hour}$$

EXAMPLE

a = distance between spikes   = 1 foot
b = distance from cross piece to eye level   = 7 feet
Z = height of cloud   = 4,000 feet
t = time for cloud to move between two spikes = 12 seconds

$$V = \frac{30}{44} \times \frac{1}{7} \times \frac{4,000}{12} = 32 \text{ miles per hour}$$

A convenient ratio for the distances a and b is 1 to 6.81. This makes the value

$$\frac{30}{44} \times \frac{1}{6.81} = 0.1$$

The factor 0.1 or $\frac{1}{10}$ may be substituted to obtain

$$V = \frac{1}{10} \frac{Z}{t}$$

if Z is reported in feet. Multiply the time in seconds by 10 and divide into the cloud height in feet. If $Z = 4{,}000$ feet and $t = 10$ seconds

$$V = \frac{4{,}000}{10 \times 10} = 40 \text{ miles per hour}$$

A Norwegian modification of the Besson nephoscope is the Grid nephoscope (Fig. 39). In using these nephoscopes it is important

to keep the eye in a fixed position while observing the motion of the cloud. This may be done by holding a pointed stick firmly in a vertical position on the ground with the point just above eye level and between the nephoscope and the eye.

The second method of observing cloud motion is by mirrored vision of the cloud. The mirror nephoscope consists of a black circular glass mirror with directions engraved at its outer perimeter. A vertical pointer and eye piece is attached to the central column supporting the mirror. The distance and height

Fig. 39.   Grid nephoscope.

U.S.W.B. photograph.

Fig. 40.   Mirror nephoscope.

above the mirror of the eye piece may be adjusted. The directions marked off are opposite the true directions so that the reflected vision will move toward the graduation indicating the direction from which the cloud is moving. Cloud speeds may be obtained as with the Besson nephoscope as shown by Fig. 38.

The accuracy of any of the foregoing methods is limited by the changes in cloud structure which may occur during the observation period. It is not always possible to know that the same point is being followed.

Chapter IV

VISIBILITY

Definition and Uses

Visibility as used in meteorology might better have been called "visual range." As defined by the United States Weather Bureau, visibility (visual range) in a definite direction is the maximum distance that prominent suitable objects like trees or houses, located in that direction and viewed against the horizon sky, are visible to an observer of normal eyesight under existing conditions of atmosphere, light, etc.

Ordinarily only one visibility value is included in a weather report, and under nonuniform conditions a value called the *prevailing visibility* is used. Under these conditions, the prevailing visibility is the greatest value of visual range that satisfies the condition that *visual ranges equal to or greater than that value exist over at least half of the horizon.*[10]

Visibility is used in two ways: (1) the pilot or navigator of an aircraft wishes to know how far he can see so that he may be able to identify objects along the flight path; (2) the meteorologist or forecaster uses the visibility or the transparency of the atmosphere as an indication of the distribution of haze, dust, smoke, or other impurities in the air.

Nature of Visibility [4, 21]

Visibility is one of the most complicated of all meteorological elements. A number of visibility instruments have been developed, but none has been considered sufficiently practical to replace observations made by the unaided eye.

When we measure visibility we are dealing with the turbidity of the atmosphere. Turbidity is a measure of the lack of transparency of the atmosphere, depending on the amount of foreign particles in the atmosphere. The turbidity may affect visibility as follows:

(*a*) The strength of light traveling through the atmosphere is reduced by absorption and scattering by the air molecules and by minute particles of matter suspended in the air. These particles are of numerous kinds: spores, bacteria, dust, smoke, fog, and ice crystals.

(*b*) The amount by which the light is lost depends upon the kind, sizes, and quantity of suspended particles. The color of light which is most scattered and that which is most transmitted depends upon the size, kind, and color of particles in the air at any particular time.

(*c*) Air molecules are of such a size that blue light is scattered about 10 times more than red light. The sun at sunset is red for this reason. The sunlight at sunset must travel through a long path of air. Along this path, the blue portion of the sunlight is scattered in all directions, so that mostly the red light is left by the time the direct light reaches us at the end of the path.

(*d*) Haze, composed of extremely small salt or dust particles, also scatters blue light more than red, though these colors do not appear as "pure" in haze as they do when the light is scattered by air molecules.

(*e*) The size of fog droplets can vary over a large range; and, since the color of light that is most scattered depends upon the size of the particles, we may expect that different types of fog will scatter colors in different degrees.

(*f*) Clean, pure fog droplets smaller than 1 micron in radius scatter blue light more than red, just as haze particles do. If the droplets are between about 1.5 and 5 microns in radius, however, they scatter *red* light more than blue. Droplets of more than 5-micron radius (the most common fog) scatter all colors about equally. (One micron equals 0.001 millimeter, equals 0.00004 inch.)

(*g*) The amount of attenuation of light thus depends upon the amount of scattering and absorption of light by the air molecules and foreign particles in the air. A measure of this ability to scatter and absorb light is called the extinction coefficient.

It is evident that visibility is related to the extinction coefficient. Visibility is inversely proportional to the amount of light absorbed and scattered. Although visibility also depends upon a number of other factors, let us attempt to understand why an object may be visible or invisible, depending upon its distance from us.

DAYLIGHT VISIBILITY

For an object to be visible, we must be able to distinguish it from its surroundings. In other words, the object must be contrasted with its background. If no contrast were perceptible, we would not be able to distinguish the object from its surroundings and it would be "invisible." This is one of the principles applied in camouflage.

For purposes of discussing daylight visibility, the principal contrast that enables us to distinguish an object is the contrast in brightness. Brightness is the luminous intensity per unit area normal (perpendicular) to the line of sight. The luminous intensity is usually expressed in candlepower, which represents a certain definite rate of flow of light energy. If the difference in brightness between an object and a brighter background is made less and less by increasing the brightness of the object, a point is finally reached where we no longer can distinguish a difference between the two. As a matter of fact, however, there is actually still some difference, but, since our eyes are not perfect instruments, the difference is not apparent to us. Now, if the *actual* brightnesses of object and background are measured when the object is just visible, their *actual* contrast can be found. The contrast is calculated on a relative basis by dividing the *difference* in brightness by the brightness of the background. This contrast is called the "threshold of brightness contrast." It is obvious, therefore, that, since all people's eyes are not the same, this value varies from person to person, within certain limits, and depends also on the adaptation of the eyes to existing light conditions.

The threshold of brightness contrast is related to visibility in the following way: Suppose that we look at a black object in the daylight against a completely cloudless sky. As we move the object farther and farther away from us, toward the horizon, it appears to become lighter and lighter, its apparent brightness becoming more and more the same as that of the sky. Finally, at some certain distance, the object becomes no longer visible, because we can discern no contrast between it and the sky.

The reason why the object appears lighter as its distance from the observer increases is easy to understand. Sunbeams from the window, slanting across a room, are made visible by the dust particles in the air. The path seems "lighted up." In the same way, though not so spectacularly, since the light is not broken up into separate beams outdoors (except in crepuscular rays), the air between an observer and an object is "lighted up" as the result of scattering. The longer the air path, the more "air-light" there is between the observer and the object. Thus, the farther away the object, the brighter it appears.

It is obvious that, by moving our black object farther and farther toward the horizon, the contrast will diminish until it has the value of the "threshold brightness contrast." At this point the object will just be visible. This distance from the observer to the object is the "visibility."

This certain distance at which the object becomes visible is related to the extinction coefficient, which becomes greater as the distance becomes less. As the number of foreign particles in the atmosphere (in the form of smoke, dust, fog, etc.), increases, the value of the extinction coefficient increases, and the visibility decreases.

Factors Which Influence Visibility [4, 21]

The concept of visibility implies a visible object. It is in the choice of a reference object that our difficulties multiply. We find that visibility depends (apart from individual differences in visual acuity) upon various properties of the object, surroundings, and lighting. These are, principally:

Reflecting power and color of object.
Reflecting power of background.
Amount of cloudiness.
Position of sun.
Angular size of object.
Nature of terrain between object and observer.
Bright light points in the field of view.

Let us consider these factors:

The Ideal Object and Background

It is found that the visibility of a black object against a sky background (with either a cloudless sky or a completely overcast sky) can easily be computed by knowing the extinction coefficient and the "threshold" values. The position of the sun in this instance makes no difference.

Effect of Different Objects and Backgrounds

To make the above ideas concrete, let us take some numerical examples. In Table IV–1 it is assumed that the same atmospheric conditions prevail, that the sky is uniformly overcast, and that the objects are of the same angular size.

TABLE IV–1

Object	Background	Visibility, miles
Black	Sky	3.0
Gray (reflecting 15% light)	Sky	2.9
White	Sky	2.5
Black	Snow	2.3
Gray (reflecting 15% light)	Snow	2.1
Black	Ground	0.9

This table shows that: (1) gray objects may be used, but they should be dark; the darker the better; (2) white objects should not be used; (3) the sky should be the background for the objects.

Position of Sun

The above points are further emphasized when we consider the effect of the position of the sun relative to the observer and object.

TABLE IV–2

		Visibility, miles		
Object	Background	Sun before Observer	Sun Directly to Right or Left of Observer	Sun behind Observer
Black	Sky	3.00	3.00	3.00
Gray (reflecting 15% light)	Sky	2.97	2.97	2.43
Black	Ground	0.72	1.11	2.91
Gray (reflecting 15% light)	Ground	0.36	0.63	2.22

Table IV–2 will illustrate this effect. The sky is assumed to be clear, and the sun is at an elevation of 20°.

Color of Objects

It has already been stated that if the objects are colored the situation becomes complex. It can be shown, however, that, if the colored objects are not *too* bright, they will behave about the same as gray objects. This fact is significant, for we see that dark-colored buildings (dark red brick, etc.) may be used as reference objects, always keeping in mind what has been said about the background, together with further considerations as to size, which will be discussed later.

As the distance of a colored object approaches the limit for visibility its color seems to disappear and it looks gray, so it may be treated as a gray object. Table IV–3 shows the visibility under the same atmospheric conditions, but using objects of different colors as reference points. It illustrates the fact that the colored object acts nearly the same as a gray object with the same reflecting power.

TABLE IV–3

Color	Reflecting Power, %	Visibility, miles
Gray	13	9.18
Red	13	9.10
Gray	30	8.95
Blue-green	30	9.05

Angular Size of Objects [4]

The angular size is the angle subtended by the object at the observer's eye. It is found that, for objects subtending angles less than 1°, the visibility decreases rapidly as the angle becomes less

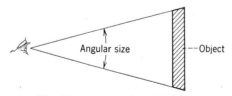

Fig. 41. Angular size of objects.

and less. Above 1° to about 5° there is very little variation, as is shown in Fig. 42, in which the angle subtended by the object is plotted against visibility.

The graph, Fig. 42, shows, for instance, that, if an object subtending 3° is just visible at 3 miles, then, under the same condi-

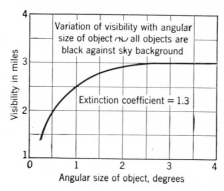

FIG. 42. Angular size of objects vs. visibility in miles.

tions, an object subtending only 0.3° could not be seen beyond 1.6 miles.

Effect of a Bright Light in the Field of View

The dazzling effect of looking toward a bright light is a common experience. If such a "dazzle source" is close to the line of sight, the object being sighted on is more difficult to see. The visual range will appear less than it actually is, because the threshold brightness contrast increases owing to the dazzling effect in the eye.

Observers unconsciously correct somewhat for this by shading their eyes with their hands when determining visibility in directions near the sun.

Effect of Terrain between Observer and Object

If the terrain between the observer and the object is highly reflecting (such as lakes, white sand, or snow) there will be an appreciable addition of "air-light" in the space between observer and object, causing the object to be less visible, of course, than it would be if the intervening terrain were dark soil or grass.

Effects of Moonlight, Starlight, and Twilight

It is well known that the visual range of objects in the subdued light of moonlight, starlight, or twilight is not so great as in daylight. The reason is obvious. As an example of the magnitude of this effect, suppose that the visibility were such as to permit an object to be just visible at 2½ miles in daylight; in bright moonlight the object would not be visible at more than about 0.6 mile.

In twilight the same sort of effect is noticeable. However, in twilight the illumination and the brightness of the sky vary with direction, so that the visibility will depend on the direction of the object.

Visibility at Night

Visibility at night, of necessity, is determined mostly by the visual range of lights rather than by objects. Therefore, only unfocused lamps should be used to determine visibility, since focused beams and beacons can naturally be seen farther than ordinary lights. It is evident that, since the distance at which lights can be seen depends upon their candlepower, this distance has no relation to the visibility determined by objects in the daytime. Candlepower is the amount of light per unit spherical angle. If the *proper candlepower* lamp is chosen, however, there will be a 1:1 correspondence between the visibility determined in daylight and the visibility determined at night by means of the lamp. Thus, under similar conditions, when the visibility in daylight is 3 miles, it would take a 225-candlepower lamp to be just visible at 3 miles at night.

In the discussions which follow, for the sake of clearness, we will arbitrarily define "visual range" as the distance at which a dark object subtending 2°–3° is just visible against the horizon sky, during daylight. The distance at which an electric lamp is just visible at night will be called "lamp range." The term "visibility" will remain as defined at the start of the chapter.

Variation of Lamp Range with Candlepower [10, 21]

The candlepower is not only dependent upon the design of the lamp and the direction in which it is turned but also upon the voltage impressed upon it.

For instance, suppose that the visual range were 3 miles; then a 60-candlepower lamp would be just visible at 2.3 miles at night (lamp range).

Now suppose that the line voltage of 110 volts is reduced by 18 per cent. The lamp would now be giving only about 31 candlepower. The lamp range would be 2 miles.

If the line voltage had *increased* by 18 per cent, the lamp would give about 106 candlepower. Now the lamp range would be 2.6 miles.

In the foregoing example, the transparency of the atmosphere was held constant, so that the visual range would have been 3 miles. Suppose now that the atmosphere becomes clearer, giving a visual range of 10 miles. What would be the behavior of our lamp range?

At 60 candlepower, the lamp would be just visible at 4.6 miles. If the voltage to the lamp were reduced by 18 per cent, the lamp would then be visible at no greater distance than 3.8 miles. If the voltage increased by 18 per cent, the lamp range would be 5.2 miles.

Not only, therefore, are most lamp ranges different from the visual range under the same atmospheric conditions, but changes in line voltage cause changes in lamp range. Thus, considering only lamps of ordinary candlepower (about 60 cp.): for poor visual range, changes in line voltage have only small (perhaps negligible) effects; for good visual ranges, 10 miles, the effect is greater. Also, for poor visual ranges the difference between visual range and lamp range is less than for good visual range.

Suppose, however, that we use a much stronger lamp, say one of 750 candlepower, at a voltage of 110. If the visual range were 10 miles, this lamp would be (under the same atmospheric conditions) just visible at night at 8 miles. If the voltage dropped by 18 per cent, the lamp range would be about 7 miles. If the voltage increased by 18 per cent, the lamp range would be about 9 miles.

Now if the atmospheric conditions were such that the visual range is ¾ mile, and a 750-candlepower lamp at 110 volts were used, the range would be about 1.4 miles. If the voltage dropped by 18 per cent, the lamp range would be only about 1.2 miles. If the voltage increased by 18 per cent, lamp range would be about 1.4 miles.

It is thus evident that, if a system of lamps is to be used for determining lamp ranges, and their voltage is not kept constant, lamps of high candlepower are better than lamps of low candlepower.

The following table shows the approximate relationships between the various factors already discussed.

TABLE IV–4

(Distances rounded to nearest tenth of a mile)

Visual Range, miles	Candlepower at 110 volts = 60 candlepower			Candlepower at 110 volts = 100 candlepower			Candlepower at 110 volts = 750 candlepower		
	Distance Lamp Is Visible at Voltages of			Distance Lamp Is Visible at Voltages of			Distance Lamp Is Visible at Voltages of		
	90	110	130	90	110	130	90	110	130
¾	0.9	0.9	1.0	0.9	1.0	1.1	1.2	1.4	1.4
3	2.0	2.3	2.6	2.2	2.6	2.9	3.3	3.6	4.0
6	3.0	3.5	4.0	3.4	4.0	4.4	5.2	5.8	6.4
10	3.8	4.6	5.2	4.4	5.2	6.0	7.0	8.1	9.0

If it were desired to relate lamp range to visual range, a factor table could easily be computed for any system of lamps.

Flashing Lights

The visual range of flashing lights may, for practical purposes, be considered the same as that for steady light, as long as the duration of each flash is several seconds. If, however, the flashes last for much less than 0.2 second, the light will not be visible as far as a steady light. As an example, suppose that a certain steady light is visible at 3½ miles; then if the same light were to flash on and off so that each flash on lasts only 0.02 second, the light would be visible at only 2 miles. Since flashing lights of such small intervals are rarely used in practice, this consideration may be ignored with respect to visibility determinations made by weather observers.

Effect of the Brightness of the Background of Lamps

If a white lamp is seen against a background that is not black but has a certain brightness, it is known that the lamp cannot be seen as far as it could be if it were against a black background. To illustrate this, let us assume the following: The visual range is 3 miles. A lamp of 225 candlepower, against a black background, can be seen at 3 miles. A similar lamp, against a background of moonlit snow, would have to be at about 2.8 miles to be

visible. Evidently, then, at least for lower visibilities, the effect
is not very important. If we examine the situation for higher
visibilities (say 10 miles), we find that a lamp which can be seen
at 8 miles against a black background would have to be moved to
about 7.6 miles to be visible against moonlit snow. Again it
appears that the difference in the visual range of the lamp is not
of too great magnitude. For red lamps, the background brightness
(within reasonable limits) has no effect.

Effect of Fog

All the preceding discussion was based on night visibility conditions of over $\frac{1}{2}$ mile and in the absence of fog.

If fog exists, the lights no longer act as point sources. The fog
particles scatter the light and form a more or less large luminous
area surrounding the lamp. The light from this luminous area is
diffused light. According to Langmuir and Westendorp ("Light
Signals in Aviation and Navigation," *Physics*, 1:273–317, 1931),
the eye is 7,000 to 170,000 times more sensitive to light from a
point source than to diffuse light, depending upon background
brightness. Therefore the diffused light caused by the fog around
a lamp cannot be seen as far as the light from the same lamp not
surrounded by fog.

As an illustration of the relative sensitivity of the eye to point
and diffuse sources, consider this: Compare the candlepower
necessary to make visible at 6 miles, first, a flash from a point
source (a lamp), and then, second, light reflected from a cloud.

In the first case, if a candlepower of 250 is sufficient, then nearly
3,250,000 candlepower will be necessary in the second case.

Effect of Color

The effect of color of a light source on visibility is substantially
the same in fog as in clear air. Red lights, including neon, of the
same candlepower as white lights cannot be seen any farther than
the white lights in a fog. For this reason it is satisfactory to use
red obstruction lights as visibility markers.

PREPARING A VISIBILITY CHART

The point of observations should be chosen so that a view of
the entire horizon may be had. If this is impossible, sufficient
vantage points near the observatory may be selected to enable an
observer to view the horizon in all directions.

Suitable objects should be chosen in all directions and at varying distances in accordance with the previous sections of this chapter. On a map of the surrounding area and with the observatory as the center concentric circles should be drawn and labeled in terms of their distances from the observatory. The various objects should also be listed so that they may be recognized.

PROCEDURE FOR DETERMINING PREVAILING VISIBILITY

Under nonuniform visibility conditions a representative value may be obtained as follows: Divide the circle of the horizon into a suitable odd number (such as 3, 5, or 7) of equal sectors. The sectors should be chosen so that the visual range for each sector is approximately uniform and can be represented by a single value. Write down the value of the visual range for each sector, and arrange the values in the order of increasing numerical value. The value that lies in the middle of the series will be the *prevailing visibility*.

Many observatories are so located that smoke or fog may be prevalent in some particular direction. It is then desirable to report the visibility in each direction. For example: N 2 miles, E 1½ miles, S ½ mile, W 1½ miles.

ESTIMATING VISIBILITY [10]

The observatory may be so located that the farthest visible objects on all horizons are so close that true visual range may not be obtained for conditions of excellent visibility. Under these conditions, the observer should estimate the visibility by noting the transparency of the atmosphere. Small objects should be observed, and the sharpness with which they stand out may serve as a guide for selecting the visibility. When the more distant objects stand out sharply with little blurring of color, the air may be considered to be free of haze and the visibility quite high. If objects are blurred or indistinct and seem to have a gray or purplish hue, the presence of haze or other obstructions and a reduced visibility are indicated. As was indicated on page 71, the size of the object is important in estimating low visibilities. Small objects should be chosen for low visibility conditions.

Visibility and Aviation

One of the most important factors in the safety of aviation is good visibility. A pilot may navigate "blind" through the course of his trip, but apparatus which will permit him to land through a region of zero visibility has not, as yet, been put into universal use, although considerable progress has been made with experimental installations. Visibility conditions must therefore be accurately forecast and observed so that the pilot may know actual conditions to expect.

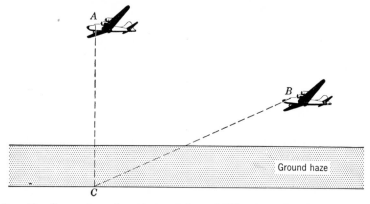

FIG. 43. Comparison of vertical and slant visibility from aircraft at different altitudes through a ground haze layer.

Meteorological reports on visibility will include only horizontal visibility from the surface. This is of greatest importance in landing and take-off but must be supplemented by a knowledge of conditions as viewed from above, especially if the pilot is relying on contact flying as the principal means of navigation, or if bombing, photography, or observation is involved.

The surface air may occasionally be relatively free from visual obscuration while a layer of haze aloft will prevent the pilot from viewing the ground. At other times, with a shallow layer of fog, the pilot may be able to view the ground quite clearly from above but finds that visibility decreases markedly when coming in for a landing. In Fig. 43 it will be seen that airplane *A* will have a much clearer line of vision to point *C* than airplane *B*, which is at a lower altitude.

Here the visibility from the ground is of major importance, rather than the visual range of the pilot in the air. It is possible to see a greater surface area by climbing; in fact, the range of visibility will be directly proportional to the height, as shown in Fig. 44. At point *B* aloft, the pilot can see from points *A* to *C*.

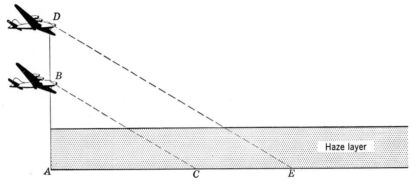

Fig. 44. Visual range from aircraft at different altitudes with a haze layer near the ground.

By climbing higher to elevation *D*, he will be able to see from *A* to *E*. It will be seen that the distance *CE* is directly proportional to *BD*. It must be assumed, however, that the air above the haze layer is relatively clear and free from pollution in both cases.

The principal use of ground observers' visibility reports is in landing. On the final glide the aircraft enters the lower layer of air through which the visibility is being determined, and the pilot's visibility becomes approximately the same as the ground observer's visibility.

Chapter V

HYDROMETEORS

INTERNATIONAL DEFINITIONS

International meteorology requires that each meteorological phenomenon be represented by a single symbol which has the same meaning in all countries. It is, therefore, necessary that the description of the phenomenon represented by a symbol be the same in all countries. It is also desirable that a name or term used to designate a given phenomenon be as widely understood as practicable without ambiguity. This last requirement is not always realized. In this text we shall follow closely the United States Weather Bureau's "Definitions of Hydrometeors and Other Atmospheric Phenomena" which was adopted November 18, 1938.

Certain names have been adopted in England and the United States to designate different phenomena. So far as practicable, it is desirable to discontinue the use of such names in meteorological literature and records. Those terms which in American usage have not been clearly defined in the past or which have been applied to different phenomena contrary to international usage are thus objectionable. A term in this category is *mist*. It has been employed in the past in the United States generally to designate *drizzle* and occasionally *light rain*, as defined in this chapter. In England the term *mist* refers to *thin fog*, similar to *damp haze* as defined here, and similar to *light fog*, but possibly somewhat more attenuated at times. To avoid the possibility of confusion the terms *mist* and *misting* are not used in United States Weather Bureau records.

Another term which has different meanings in England and North America is *sleet*. In North America *sleet* generally refers to *ice pellets*, i.e., frozen raindrops, or ice pellets formed by freezing of melted or partially melted snowflakes. In England *sleet* refers to *rain and snow mixed*. Some people in North America also use the term *sleet* in this sense. In United States Weather Bureau records *sleet* is the name applied to ice pellets.

Visibility and Intensity

Visibility is used as a criterion for estimating the degree of intensity of certain (not all) phenomena among which may be mentioned fog, ice fog, smoke, dust and dust storm, sand and sand storm, drifting and blowing snow, snow, drizzle, and exceptional visibility.

Restrictions to visibility produced by particles not falling from aloft out of clouds may be of two types: (*a*) lithometeors; (*b*) hydrometeors.

Obstructions to Vision (Not Including Precipitation)

Dry Haze ∞

Dust or salt particles which are dry and so extremely small that they cannot be felt or discovered individually by the unaided eye; however, they diminish the visibility and give a characteristic smoky (hazy and opalescent) appearance to the air.

These lithometeors produce a uniform veil over the landscape and subdue its colors. The veil has a bluish tinge when viewed against a dark background such as a mountain, but it has a dirty yellow or orange tinge against a bright background, such as the sun, clouds at the horizon, or snow-capped mountain peaks. It is thus distinguished from grayish = (light fog), the thickness of which it may sometimes attain. When the sun is well up, its light may have a peculiar silvery tinge due to haze.

Irregular differences in air temperature may cause a shimmering veil over the landscape, and this is called "optical haze."

Damp Haze ∞

Microscopically small water droplets or very hygroscopic particles suspended in the atmosphere, but the horizontal range of visibility is 1¼ miles or more, usually considerably more. Similar to a very thin fog, but the droplets or particles are more scattered than in = (light fog), and presumably also smaller.

This hydrometeor is usually distinguished from dry haze (see "haze" above) by its grayish color, the "greasy" appearance of clouds seen through "damp haze" as though viewed through a dirty window pane, and the generally high relative humidity. It is commonly observed on seacoasts, and in southern states, most frequently with onshore winds and in the vicinity of tropical disturbances. A common mode of formation of "damp haze" is the carrying up to high levels of particles from salt-water spray in windy weather. In contrast, light fog is more commonly observed when there is little movement of the surface air.

Light Fog =

Microscopically small (minute) water droplets or very hygroscopic particles, too small to be seen individually by the unaided eye, suspended in the atmosphere. The horizontal range of visibility is ⅝ mile (3,300 feet), or more, because the minute water droplets are much smaller and more scattered than in ≡ (fog).

This hydrometeor is similar to ≡ (fog), except that the droplets are smaller and more scattered so that the horizontal visibility is not reduced to less than ⅝ mile (3,300 feet). At temperatures above the freezing point, the relative humidity in = (light fog) generally ranges from 90 to 97 per cent. Light fog = always looks more or less grayish.

Fog ≡

Minute water droplets suspended in the atmosphere, reducing the horizontal range of visibility to less than ⅝ mile (3,300 feet).

If fog does not obscure the sky in all directions, record the sky conditions and add notes regarding distribution of the fog, such as ≡ SE, etc.

This hydrometeor is easily distinguished from haze, dust, or smoke by its essential wetness. At temperatures above the freezing point true fog is characteristically accompanied by a relative humidity greater than 97 per cent. Consequently the air feels clammy and humid when fog is present, and on close examination one may even see the individual droplets floating past the eye. True fog generally looks whitish, except in the vicinity of industrial regions where it may be a dirty yellow or gray. The symbol ≡ (fog) is to be recorded only when the fog reduces the horizontal visibility to less than ⅝ mile (3,300 feet). Fog will always be recorded as at observational elevation. Proper notations should be recorded as to whether different conditions prevail at other elevations. When fog and smoke or any mixture of fog and dust, etc., occurs simultaneously record as "≡," or according to proper lithometeor.

Intensities

Moderate	When fog is present and the visibility is $\frac{5}{16}$ mile (1,650 feet) to (but not including) $\frac{5}{8}$ mile (3,300 feet).
Heavy	When fog is present and the visibility is less than $\frac{5}{16}$ mile (1,650 feet).

Ice Fog ⇔

The occurrence of fog in the form of ice crystals or spicules. This phenomenon occurs more frequently in higher latitudes. Sun is usually visible, but horizontal visibility is considerably restricted. It is colloquially termed "frost in air," "frozen fog," which in some western parts of the United States and in Alaska is called "Pogonip."

Intensities

Light	When ice fog is present and visibility is $\frac{5}{8}$ mile (3,300 feet) or more (see Ice crystals).
Moderate	When ice fog is present and the visibility is $\frac{5}{16}$ mile (1,650 feet) to (but not including) $\frac{5}{8}$ mile (3,300 feet).
Heavy	When ice fog is present and the visibility is less than $\frac{5}{16}$ mile (1,650 feet).

Smoke ◯

The presence in the air of particles of foreign matter resulting from combustion. In light amounts it may be confused with haze or fog but usually can be differentiated from them by its odor.

Sun's disk at sunrise and sunset is very red and during daytime has a reddish tinge. Smoke from a distance, as from forest fires, usually is of a light grayish or bluish color and evenly distributed in the upper air. City smoke may be brown, dark gray, or black.

Intensities

Light	When smoke is present and visibility is $\frac{5}{8}$ mile (3,300 feet) or more.
Moderate	When smoke is present and the visibility is $\frac{5}{16}$ mile (1,650 feet) to (but not including) $\frac{5}{8}$ mile (3,300 feet).
Heavy	When smoke is present and the visibility is less than $\frac{1}{16}$ mile (1,650 feet).

Dust ⚲

Dust in the air to an extent that diminishes the visibility, but not less than ⅝ mile (3,300 feet).

Dust in upper air from great distances may give a grayish appearance to the sky and reduce its blueness. Occasionally the dust aloft may be of a brownish or yellowish hue. Depositions of dust may be observed on shiny surfaces, and, when rain or drizzle occurs, a muddy residue on car finishes, windows, etc., may be noticed.

Intensities

Light	When dust is present and visibility is ⅝ mile (3,300 feet) or more.
Moderate	When dust is present and the visibility is ⁵⁄₁₆ mile (1,650 feet) to (but not including) ⅝ mile (3,300 feet).

Drifting Snow ⤲

Snow raised from the ground and carried by the wind, so that the horizontal visibility becomes less than ⅝ mile (3,300 feet), although no real precipitation is falling. The snow is drifting so low above the ground that the vertical visibility is not appreciably diminished.

Intensities

Light	When drifting snow is present and visibility is ⅝ mile (3,300 feet) or more.
Moderate *	When drifting snow is present and the visibility is ⁵⁄₁₆ mile (1,650 feet) to (but not including) ⅝ mile (3,300 feet).
Heavy *	When drifting snow is present and the visibility is less than ⁵⁄₁₆ mile (1,650 feet).

* Intensities referred to as "moderate" or "heavy" conform to international definition which limits the visibility to ⅝ mile (3,300 feet) for these phenomena; the "light" is added for greater visibilities.

Blowing Snow ⤲

Snow raised from the ground and carried up by the wind so that the horizontal visibility becomes less than ⅝ mile (3,300 feet), although no real precipitation is falling. The snow is carried so high up from the ground that the vertical visibility is reduced considerably.

If it cannot be distinguished whether or not precipitation of snow is also occurring, use the symbol ✳✝, to represent what is commonly called "blizzard."

Intensities

Light   When blowing snow is present and visibility is ⅝ mile (3,300 feet) or more.

Moderate *   When blowing snow is present and the visibility is $\frac{5}{16}$ mile (1,650 feet) to (but not including) ⅝ mile (3,300 feet).

Heavy *   When blowing snow is present and the visibility is less than $\frac{5}{16}$ mile (1,650 feet).

* Intensities referred to as "moderate" or "heavy" conform to international definition which limits the visibility to ⅝ mile (3,300 feet) for these phenomena; the "light" is added for greater visibilities.

Precipitation Forms

Precipitation is composed of particles falling from aloft out of clouds. Three general classes of hydrometeors may be listed, depending upon the nature of the precipitation.

(*a*) Hydrometeors which may fall more or less continuously, or in showers: rain, snow, sleet.

(*b*) Hydrometeors which fall almost exclusively in showers: snow pellets, small hail, hail.

(*c*) Hydrometeors which never fall in showers: drizzle, snow grains, ice crystals.

Precipitation—Continuous or Shower

Rain ●

Precipitation of drops of liquid water in which the drops are generally larger than 0.02 inch in diameter; they fall faster than 10 feet per second in still air. (See drizzle.)

This hydrometeor ● (rain) must be carefully distinguished from ❥ (drizzle). While the drops in rain are generally larger than 0.02 inch and those in drizzle always smaller than 0.02 inch, rain drops may, under certain circumstances, have diameters also smaller than 0.02 inch. Such rain drops are much sparser than drizzle drops. The first rain drops from an advancing altostratus-nimbostratus canopy of clouds may all have diameters less than

0.02 inch, but then they can be distinguished from drizzle by the clouds from which they fall, by not being so numerous, and by not materially reducing the previously existing horizontal visibility as drizzle does. Precipitation of rain may be either fairly continuous or showery, but that of drizzle is rather uniformly continuous and never showery (see showers, for which a special symbol is used). Thus, rain can occur from shower clouds, such as cumulonimbus or cumulus, but drizzle does not. When the precipitation is rather continuous, a consideration of the clouds is helpful in avoiding confusion between the two phenomena in question. Rain, as distinguished from showers of rain, manifests itself as fairly continuous precipitation of ordinary rain drops from a continuous sheet of cloud. The sky is, as a rule, covered with a layer of real rain clouds (i.e., nonshower clouds such as nimbostratus) formed by progressive lowering from an altostratus system, or with a uniformly gray but relatively high canopy of clouds (altostratus), generally with formless masses of cloud below (scud: fractocumulus or fractostratus), which may even be present in such quantities that the upper clouds are completely hidden. On the other hand, drizzle usually occurs from a continuous dense and low layer of stratified cloud which has formed near the surface, but not from progressive lowering of higher clouds. (See drizzle.)

Intensities

Light   The rate of fall at the time of observation lies within and includes the following limits: Trace per hour to 0.10 inch per hour (0.01 inch or less in 6 minutes).

Moderate   The rate of fall at the time of observation lies within and includes the following limits: 0.11 inch per hour to 0.30 inch per hour (more than 0.01 inch to 0.03 inch in 6 minutes).

Heavy   The rate of fall at the time of observation is in excess of 0.30 inch per hour (over 0.03 inch in 6 minutes). (The above do not refer to accumulated depth of rain over a period of hours.)

Snow ❋

Precipitation of water (in the solid state) mainly in the form of branched hexagonal crystals or "stars" (snowflakes), often mixed with simple ice crystals.

When a considerable number of snowflakes and detached, simple ice crystals fall together use the symbol $\overset{\leftrightarrow}{\underset{*}{}}$. In extremely cold weather in high latitudes precipitation may occur exclusively in the form of detached, simple ice crystals. When this occurs a special note must be made by the observer. To be distinguished from $\overset{\leftrightarrow}{\ominus}$ (ice fog) and ↔ (ice crystals).

This hydrometeor ordinarily manifests itself in the form of single platelike crystals of water in the solid state, or clusters of such crystals (flakes) which fall more or less continuously from a solid cloud sheet.

Because of the mechanical difficulties involved in measuring snow as a rate of weight-accumulation, visibility is used as the criterion for determining the intensity of snowfall (see below).

Intensities

Light   When falling snow is present and visibility is ⅝ mile (3,300 feet) or more.

Moderate   When falling snow is present and the visibility is $\frac{5}{16}$ mile (1,650 feet) to (but not including) ⅝ mile (3,300 feet).

Heavy   When falling snow is present and the visibility is less than $\frac{5}{16}$ mile (1,650 feet).

Rain and Snow, Mixed ⨳

Precipitation of ● (rain) and ✳ (snow) together.

Record "Rain and snow mixed" if rain predominates; record "Snow and rain mixed" if snow predominates.

The observer will be guided by his experience and observation in recording light, moderate, or heavy rain and snow mixed ⨳. When rain and snow are occurring alone, on as many occasions as possible he must correlate the visibility with his sense of the rate of fall for each of the three intensities. He will thus by experience obtain an impression of the rate of fall of snow alone as well as of rain alone which pertain to the respective intensities: light, moderate, and heavy. When "rain and snow mixed" occurs, the visibility criterion for snow intensity will be abandoned and the observer will estimate and record the intensity of the rain and of the snow, separately, by judgment on the basis of his experience with apparent rates of fall of each.

Sleet △

Transparent or translucent globular or irregular hard pellets of ice, about 0.04–0.16 inch in diameter; they rebound when falling on hard ground.

The formation of this hydrometeor occurs only when ordinary rain drops or partially melted snowflakes freeze while falling through a surface layer of relatively cold air, all or part of which is characterized by temperatures considerably below freezing. Thus, the temperature at ground level is almost always below freezing when △ (sleet) is observed. The sky is covered with a solid layer of cloud which helps to distinguish △ (sleet) from △ (small hail).

Intensities

Light	The falling of ice pellets is restricted to a few scattered pellets.
Moderate	The falling of ice pellets in moderate numbers so as to cause some accumulation on the ground.
Heavy	Falling of ice pellets in such considerable numbers as to cause rapid accumulation.

SHOWER TYPES OF PRECIPITATION

Snow Pellets ⚹

White, opaque, or round (seldom conical) grains of snowlike structure, about 0.02–0.20 inch in diameter in all directions. They are crisp and easily compressible; they rebound when falling on hard ground, and thereby often burst.

The occurrence of this hydrometeor is restricted chiefly to occasions when the surface temperature is at or slightly below the freezing point. It often falls before or together with ❋ (snow). If its occurrence has a showery character it should be recorded by means of the symbol ⚹̷. It is sometimes called "graupel" or "tapioca snow."

Intensities: Same intensities as for sleet.

Small Hail △

Semitransparent, round or sometimes conical grains of frozen water, about 0.08–0.20 inch in diameter. Generally each grain has a ⚹ (snow pellet) as a nucleus, with a very thin ice layer around it, which gives the grains a glazed appearance. They are not easily compressible or crisp, and, even when falling on hard ground, they ordinarily do not rebound or burst, as they are generally wet.

This hydrometeor falls from broken shower clouds, which helps to distinguish it from △ (sleet). It is the type of hail which characteristically occurs in winter and early spring along the Pacific Coast of the United States. During its occurrence the temperature is nearly always above the freezing point. It often falls together with ● (rain).

Intensities

Light	Fall restricted to a few scattered grains.
Moderate	Fall is sufficient to cause a medium rate of accumulation on the ground.
Heavy	Fall is such as to cause rapid accumulation on the ground.

Courtesy of the U.S.W.B.

Fig. 45. Comparison of hen egg and hail. Hail storm at Girard, Illinois, August 13, 1929.

Hail ▲

Ice balls, or broken portions of balls, the diameter of which ranges from 0.20 inch to 2 inches or larger, which either fall detached or in irregular lumps. On the ground they may fuse into irregular lumps. They are either largely transparent or composed of alternating clear and opaque, snowlike layers, the clear layers being at least 0.04 inch in thickness.

This hydrometeor is distinguished from △ (small hail) not only by difference in size and structure but also by the fact that it falls almost exclusively in thunderstorms and when the surface temperature is considerably above freezing. Ordinarily occurs with rain, in which event the symbol would be ▲̇, or as ▲̇ showers.

Intensities

Light   Fall restricted to a few scattered stones.
Moderate   Fall is sufficient to cause a medium rate of accumulation on the ground.
Heavy   Fall is such as to cause rapid accumulation on the ground.

PRECIPITATION WHICH DOES NOT FALL IN SHOWERS

Drizzle ❜

Rather uniform precipitation consisting exclusively of minute and very numerous drops of water (diameter less than 0.02 inch) which seem almost to float in the air, and thereby visibly follow even slight motion of the air. This must not be confused with light rain. (See rain.)

This hydrometeor usually falls out of a low, rather continuous and thick layer of stratus clouds which may even touch the earth. In cold weather the layer may be thin. It is not to be confused with the very small but *scattered* droplets which fall at the foremost edge of a general rain area. ❜ (drizzle) does not usually occur with showery weather; when showers do occur during drizzle they are the result of instability above or below the warm, moist layer where the drizzle is forming; therefore, even the finest rain from shower clouds must not be recorded as ❜ (drizzle). Along west coasts and particularly in mountains, ❜ (drizzle) may sometimes produce considerable amount of precipitation (as much as 0.04 inch per hour). This hydrometeor is associated with ≡ (fog); in fact, the two often occur simultaneously. Therefore ❜ (drizzle) is characteristically accompanied by poor visibility, and this criterion should be used in distinguishing it from fine rain. Drizzle most commonly occurs when a relatively warm, moist layer of air flows over a cooler surface layer of air or cooler ground (or sea) surface so that condensation ensues in the moist air, with formation of low stratus clouds, often with fog. If the warm air in question is an unstable, turbulent layer of high humidity near the surface, the condensation generally takes place more extensively and the precipitation is heavier than otherwise. This also tends to be true when the motion of the moist air is upward along an upsloping cooler surface layer of air or upsloping ground.

Intensities

Light	When drizzle is present and visibility is $\frac{5}{8}$ mile (3,300 feet) or more.
Moderate	When drizzle is present and the visibility is $\frac{5}{16}$ mile (1,650 feet) to (but not including) $\frac{5}{8}$ mile (3,300 feet).
Heavy	When drizzle is present and the visibility is less than $\frac{5}{16}$ mile (1,650 feet). When drizzle occurs with visibilities less than $\frac{5}{16}$ mile (1,650 feet), fog is usually present and the phenomena would then be recorded fog and drizzle.

Ice Crystals ↔

Small or tiny unbranched ice crystals in the shape of staffs or plates, often so small that they seem to float in the air.

This hydrometeor is either an ice fog so thin that it scarcely affects the visibility or a type of active precipitation often with other forms (see note under snow). As an ice fog it is observed only during extremely cold, winter weather. It glitters in the sunshine, whence it is sometimes called "diamond dust." It often gives rise to sun pillars or to other halo phenomena. This should be distinguished from light ice fog by the greater visibility.

MISCELLANEOUS PHENOMENA

Other phenomena which may be included as part of a meteorological observation may be listed under three general headings: (*a*) water (liquid or solid) deposits on the ground or objects, formed by condensation, freezing, or sublimation; (*b*) electrical and sonic phenomena; (*c*) optical phenomena.

WATER (LIQUID AND SOLID) DEPOSITS

Dew ⌒

Water drops deposited by direct condensation of water vapor from adjacent clear air, mainly on horizontal surfaces cooled by nocturnal radiation.

When tropical maritime air blows against cold surfaces, a deposit of water drops is formed. This is called "wind dew."

Frost ⌣

Thin ice crystals in the shape of scales, needles, feathers, or fans, deposited by sublimation under conditions similar to dew formation, except that temperatures at or below 32° F. must prevail at surfaces where ⌣ is deposited. (Sublimation: To pass directly from the gaseous to the solid state without apparently liquefying.)

Intensities

Light Frost — Surface objects, vegetation, etc., covered with a thin deposit of frost which in agricultural meteorology is understood to have no destructive effect on vegetation, except to tender plants and vines.

Heavy Frost — Surface objects, vegetation, etc., covered with a copious deposit of frost which in agricultural meteorology is understood to be more destructive to vegetation than light frost, but which, however, may kill some tender staple products but not all staple products of the locality.

Killing Frost — Understood in agricultural meteorology to be a deposit of frost on surface objects, vegetation, etc., accompanied by a temperature low enough to be destructive to vegetation and staple products. It is ordinarily determined by the destructive effects on vegetation and staple crops, but, when the staple crops have passed the stage where injury is possible, the killing intensity of the frost must be determined by the heaviness of the deposit and the minimum temperature.

Sometimes temperatures below freezing and severe enough to destroy vegetation, including staple crops, occur without an actual deposit of frost crystals. (This is sometimes called "black frost.") The exact nature of the degree of severity of "heavy" and "killing" frost may not be immediately determinable, depending in some circumstances on the previous weather. If previous weather has been warm and generally favorable for growth of crops, they may be in a succulent condition and more susceptible to damage from a temperature only slightly below freezing. On the other hand, if the previous weather has favored dormancy in the spring or rapid maturity in the

fall, crops may be able to withstand lower temperatures without appreciable harm. Also, in
the drier, or semi-arid, section of the country,
lower temperatures may not cause as serious harm
as they would in more humid regions.

Soft Rime ∨

White layers of ice crystals deposited (built up) chiefly on
vertical surfaces, especially on points and edges of objects, generally in supercooled ≡ (fog) or ═ (light fog). On the windward

U.S.W.B. photograph.

Fɪɢ. 46. Rime formation on Mt. Washington Observatory.

side soft rime may grow to very thick layers or long feathery cones,
or needles pointing into the wind and having a structure similar
to that of frost. This process is probably analogous to the formation of snow pellets, in that frost crystals (formed by sublimation)

are mixed to some extent with the frozen fog or cloud droplets. Rime feathers often become impregnated with blown snow, making a more friable deposit that feels "greasy" between the fingers. (Supercooled: Cooled below the freezing point without solidification.)

Hard Rime ▼

Opaque, granular masses of ice deposited (built up) chiefly on vertical surfaces in "wet fog" at temperatures below 32° F. thus getting a more compact and amorphous structure than soft rime, analogous to that of small hail.

There are all graduations between hard and soft rime. Hard rime will build out into the wind in glazed cones or feathers.

Glaze ∽

Homogeneous, transparent ice layers which are built up on horizontal as well as on vertical surfaces either from supercooled ● (rain) or ❥ (drizzle), or from ● (rain) or ❥ (drizzle), when the surfaces are at temperatures of 32° F. or lower.

It often forms a matrix for sleet pellets that fall at the same time.

ELECTRICAL AND SONIC PHENOMENA

Thunderstorm ⟨

Thunder heard and lightning seen at the station, with not more than 10 seconds' difference of time.

A thunderstorm occurring within 2 miles of the station, regardless of whether lightning is seen, so long as thunder is heard, shall be considered in this category and shall be recorded by the symbol ⟨. Observers will be guided by the above stipulations whenever practicable. Where the time difference is not ascertainable, the observer will be guided by his judgment of the distance of the storm from the station. Sometimes the sound of the thunder itself will be the only guide, sharp thunder without rolling or rumbling indicating a thunderstorm in the close proximity of the observer.

Distant Thunderstorm ⟨⟩

Either thunder and lightning in the distance (time difference greater than 10 seconds) or thunder, but no lightning, at the station (where the thunderstorm is more than 2 miles from the station).

Observers will be guided by the stipulations given above whenever practicable. If thunder is heard, but no lightning seen, the observer will be guided by his judgment of the distance of the storm from the station. If the observer can determine that the storm is clearly 2 miles or more from the station, it should be called a distant thunderstorm.

Distant Lightning ⋖

Distant lightning without thunder audible at the station.

Tornadoes

Tornado weather may be described as sticky, sultry, and oppressive, with generally southerly winds. An hour or two before the tornado, a topsyturvy sky of clouds appears which crazily bulges down instead of up. To the west and northwest, great

Courtesy of U.S.W.B.

Fig. 47. Tornado forming and in progress. Picture at right shows tornado striking farmhouse, which appears to explode. Photograph by Mrs. Roy Homer at Gothenburg, Nebraska, autumn, 1930.

thunderstorms appear to be approaching. The towering thunderclouds have an ominous appearance—the color is often described as a sickly greenish black, and the lower clouds are generally visible in rapid and confused motion. Then, out of the base of this dark thundercloud, a ropelike funnel cloud reaches towards the earth spinning counterclockwise. The funnel varies greatly in appearance. Often it is a sort of thin, dangling rope; sometimes it resembles a gigantic "elephant's trunk"; and occasionally it resembles a fairly wide and solid-looking funnel. The air whirl around the funnel emits a roaring noise that has been compared to the roar of hundreds of airplanes in flight or scores of freight trains going through a tunnel.

The time of day most favorable for tornadoes is midafternoon to early evening, and they generally occur during the spring and summer months. They may, however, occur at any hour and during any month of the year.

Optical phenomena

Exceptional Visibility ◌

Unusual clearness and transparency of the atmosphere; distant objects and their details stand out in full relief from the background with great sharpness and distinctness, without any softening veil at 6 miles or more.

Sometimes the air is clear enough to permit prominent objects to be seen even at 100 miles.

Solar Corona or Lunar Corona ◌ *or* ◌

Luminous ring directly surrounding a luminary (sun or moon), of a radius seldom exceeding a few degrees, its inner part being bluish, whitish, or yellowish, or showing (faintly) as a whole all the spectral colors; in the last case, however, the red colors always occupy the outer part of the phenomenon.

Sometimes one may observe several repetitions of the spectral colors, or portions of them, which may be somewhat irregularly distributed, in which event it is called iridescence. Solar coronas or lunar coronas are formed by the diffraction of the light from the luminary by water droplets.

Solar Halo or Lunar Halo ⊕ *or* ◌

Luminous ring around a luminary (sun or moon) at a distance of about 22°, mostly whitish, but sometimes showing spectral colors, when the inner edge of the ring is always reddish or brownish, the other colors following in the outer parts of the ring and generally being less prominent.

Inside the ring the sky is darker than outside it. Solar halo or lunar halo is formed by refraction of the light from the luminary in ice crystals.

This description applies only to the 22° halo. The other phenomena, such as the 46° ring, parhelia, and tangent arcs, should be described in ordinary language.

Fog Bow ⌒•

A whitish, semicircular arc, or portion thereof, seen opposite the sun in fog. Its outer margin has a reddish and its inner a bluish tinge, but the middle of the band is quite white.

A supernumerary bow is sometimes seen inside the first and with the colors reversed. The bows are produced in the same way as the ordinary rainbow, but, owing to the smallness of the drops, the colors overlap and the bows appear white and the radius is smaller than that of the ordinary rainbow.

Rainbow ⌒

A semicircular arc or portion thereof seen opposite the sun in rain, usually exhibiting all the primary colors, the red being on the outside.

In addition to the internal reflection within the rain drops which gives the primary bow, a second internal reflection often produces a secondary bow outside the primary one. In this case the colors are reversed and the space between the inner and outer bows appears darker than the space inside the primary bow or outside the secondary bow.

Aurora △

A luminescence, usually seen in the northern skies in lower latitudes, but overhead, or even to the southward of the zenith, in higher latitudes. The aurora may take many forms such as arcs, rays, curtains, or coronas. It is usually whitish but may have various other colors.

The name "northern lights" is frequently given to this phenomenon. In lower latitudes it may be accompanied by a magnetic storm. The lower edges of the arcs and curtains of the aurora are usually fairly well defined, but the upper edges are ill defined and look like the brush discharge of static electricity.

HYDROMETEOR SYMBOLS

⛢	Aurora.	Popularly called "Northern Lights."
☉	Corona, Solar.	Luminous glow around the sun.
☽	Corona, Lunar.	Luminous glow around the moon.
⌓	Dew.	Condensed water on grass, stones, etc.
❵	Drizzle.	Precipitation of small water droplets.
⚇	Dust.	Presence of dust in the air.
⚇	Dust storm.	Heavy blowing dust.
=	Fog, light.	Fog with visibility ⅝ mile or more.
≡	Fog, moderate. }	
≡+	Fog, heavy. }	Fog with visibility less than ⅝ mile.
⇆	Ice fog.	Fog of ice crystals or spicules.
⌒•	Fog bow.	Arc sometimes seen in fog opposite sun, colorless.
⌐	Frost.	Ice crystals deposited on grass, stones, etc., when temperatures are 32° or lower.
∾	Glaze.	Ice layers formed from super-cooled rain or drizzle.
⊗	Haze, damp.	Like a very thin fog, visibility usually over 1¼ miles.
∞	Haze, dry.	Very fine dust in air; air seems smoky or opalescent.
▲	Hail.	Iceballs, usually fall in thunderstorms.
△	Hail, small.	Small iceballs, usually appear glazed.
⊍	Halo, lunar.	Luminous ring around the moon.
⊕	Halo, solar.	Luminous ring around the sun.
↔	Ice crystals.	Fine needles of ice floating in the air.
⟨	Lightning, distant.	Lightning in the distance, without audible thunder.
●	Rain.	Precipitation of fairly large water drops.
⁎	Rain and snow, mixed.	Precipitation of rain and snow together.
⌒	Rainbow.	Colored arc seen opposite the sun in rain.
⩒	Rime, hard.	Opaque masses of ice built up on objects.
∨	Rime, soft.	Ice crystals built up on objects.
⚇	Sand storm.	Heavy blowing sand.
▽	Showers.	Irregular falls of hydrometeors, rain, hail, etc.
△	Sleet.	Precipitation of ice pellets or frozen rain drops.
◇	Smoke.	Particles of foreign matter in the air from combustion.
⁂	Snow.	Precipitation of hexagonal crystals or flakes.
⤨	Snow, blowing.	Snow driven up in the air by the wind.
⤨	Snow, drifting.	Snow driven along the ground by the wind.
⩜	Snow grains.	Flattened or oblong grains of snow.
⤨	Snow pellets.	Small, round, crisp pellets of snow.
⎗	Thunderstorm.	Thunder and lightning in the vicinity.
◊	Visibility, exceptional.	Unusual clearness and transparency in the air.

Chapter VI

TEMPERATURE

UNITS OF TEMPERATURE

Temperature is defined as the thermal state of an object which enables it to communicate heat to other objects. Temperature changes which are effected by loss or gain of heat may be detected by changes in volume of liquid, solid, or gas, changes of state from solid to liquid, liquid to gas, or solid to gas, and change of electrical resistance.

Two basic scales of temperature are used in meteorology, the Centigrade and the Fahrenheit.

They may be compared at fixed points, such as the temperature of melting ice and the temperature of the vapor of boiling water at a pressure of 14.7 pounds per square inch (1,013.2 millibars).

TABLE VI–1

Temperature Scale	Ice Point	Boiling Point
Fahrenheit °F.	32°	212°
Centigrade °C.	0°	100°
Centigrade Absolute °A. Kelvin °K.	273°	373°
Fahrenheit Absolute °A. Rankin °R.	492°	672°
Réaumur * °R.	0°	80°

* Used in old records from some European countries.

For computations and in upper-air work the absolute scale of temperature is used. This is based on absolute zero temperature (lowest temperature physically possible) expressed in terms of the temperature scale used in computations.

99

The formula for conversion from Fahrenheit to Centigrade is

$$C° = \tfrac{5}{9}(F° - 32°)$$

METEOROLOGICAL TEMPERATURES

Almost everyone knows that air temperature near the ground is measured as a part of ordinary weather observations. When we say that it was a hot day or a cold day we are referring to the temperature of the air to which we are exposed. The meteorologist who wishes a complete description of the state of the atmosphere also requires temperature observations of the upper air. As ground or water surfaces affect the temperature of the lower air, temperatures of the soil or snow at various depths, and of water near the surface of the oceans, lakes, and rivers, are also required.

THERMOMETERS [3]

Three principal types of thermometers are used in meteorological work. They are:

Liquid	Mercury in glass
	Alcohol in glass
	Mercury in steel
Deformation	Bourdon tube
	Bimetallic
Electrical	Resistance
	Thermoelectric

Most commonly used are the direct-reading liquid types. The deformation thermometers are used principally in thermographs for recording temperature. The deformation of the bimetallic strip or Bourdon tube causes a pen to move on a graph mounted on a drum which is turned by clockwork; a continuous record of temperature may be obtained in this manner. Electrical thermometers are used for distant reading. The thermometer's unit may be exposed outside and electrically connected with indicating dials in the observatory.

SELECTING THE OBSERVATION POINT [10, 11]

For temperature measurements of the free air near the ground the site should be chosen to conform with the rules given in Chapter II. It is not desirable to measure temperature on the top of a high building. If possible, a grass plot with good exposure to the sky should be selected as the observation point for locating the instrument shelter in which the thermometer is to be mounted.

For temperature measurements of the soil or snow surface a point near (but not under) the instrument shelter may be selected for convenience. The surface must be exposed to the sky.

For sea surface temperatures taken at sea, a thermometer may be installed in the intake pipe of the ship's cooling system. Another method is to take water samples by means of a canvas bucket and to measure the temperature of the water by means of an ordinary thermometer.

Thermometer Shelter or Screen

The thermometer shelter is used to screen the thermometers against the direct radiation from the sun and objects on the earth, and to keep the thermometers dry. The shelter is a wooden box painted white with louvered sides made in such a way that air can move through it with the greatest possible freedom. The shelter should be freely exposed to sun and wind and should not be shaded by trees or buildings.

U.S.W.B. photograph.

Fig. 48. View of thermometers inside instrument shelter.

In tropical countries where permanent short grass is not available, the thermometers are hung in a wire cage within a hut which shades the ground below the instruments but allows a free circulation of air.

The thermometers should be installed at a central point within the shelter which will place them *4 feet, 6 inches,* above the ground.

The opening side of the shelter should be to the north (in the northern hemisphere) to avoid the effects of the sun showing on the instruments while the observations are being made.

The thermometer or instrument shelter may contain the following instruments:

Dry-bulb thermometer.
Wet-bulb thermometer.
Maximum thermometer.
Minimum thermometer.
Thermograph.
Hygrograph.

There should be a space of at least 3 inches between the bulbs of the thermometers and the top, bottom, or sides of the shelter. The thermometers should be arranged so that all parts of their scales may be read without moving them (except the whirling psychrometer). The maximum and minimum thermometers should be clamped down so that strong winds cannot shake them, as vibration may displace the indexes.

READING THE THERMOMETERS [10]

Degree of Accuracy Required

To obtain satisfactory values for humidity calculations the *difference* between wet- and dry-bulb temperature readings must be known to tenths of a degree Fahrenheit. The normal small-scale fluctuations in air temperature near the surface may be as great as plus or minus 1° in 1 minute. The lag of the ordinary thermometer will smooth out most of the fluctuations, and the reading will give a fairly representative value of the temperature. For air temperature values alone, the thermometer should be read to the closest degree Fahrenheit without estimating tenths.

For soil and snow surface thermometers the reading should be to the nearest tenth of a degree.

Position for Reading the Thermometer

As the mercury column and the scale of the thermometer are separated by the thickness of glass of the tube, an error, called parallax error, may be made if the eye is not level with the top of the meniscus. To avoid the error the line of sight from the eye to the top of the mercury column should be at right angles to the stem of the thermometer (Fig. 49).

The thermometer should be read as rapidly as possible in order to avoid temperature changes caused by heat from the body; also, the observer should stand as far from the thermometer as accurate reading of the scale permits.

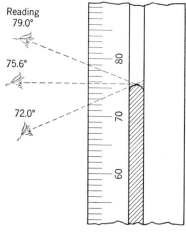

When observing by artificial light at night, care should be taken to avoid heating the thermometers with the lamp.

Estimating Tenths of Degrees

Most thermometers are marked in whole degrees, and it is necessary to estimate tenths of a degree by making imaginary divisions of the space between degree marks.

Fig. 49. Errors due to parallax in reading thermometers.

Thermometer Corrections

Thermometers used by the United States Weather Bureau are calibrated by comparison with standard thermometers, and a correction card is furnished for each thermometer.

The corrections from the card are applied algebraically; that is:

1. When the signs of the correction and of the temperature are the same, add the correction to the reading and prefix their common sign. If the correction for a temperature reading of +80° F. is +0.2° F. then the corrected temperature is +80.2° F.

2. When the signs of the correction and of the temperature readings are different, subtract the correction from the temperature reading and prefix the sign of the larger. If the temperature reading is +0.1° F. and the correction −0.2° F. then the corrected temperature is −0.1° F.

All thermometers used by the United States Weather Bureau are stem-graduated by the maker. The freezing point, 32° F. or 0° C., is found by immersing the bulb and most of the column of mercury or spirit in a bath of pure melting ice. The calibration card is prepared by placing the thermometer in a temperature-controlled bath and obtaining a comparison with a substandard thermometer. The substandard thermometer has been carefully calibrated by reference to a gas thermometer. The bath is usually alcohol cooled by dry ice (CO_2) and used for temperatures below freezing; water heated by an electrical element is used for temperatures above freezing.

CURRENT TEMPERATURE [10]

Current temperature is the temperature of the air at the present time. It is obtained by reading an exposed thermometer. For temperatures between $-30°$ F. and $+120°$ F., a mercury-in-glass thermometer is generally used. For lower temperatures an alcohol-in-glass must be substituted as mercury freezes at $-38°$ F. Other liquids are also suitable, such as pentane and toluol. All these liquids have expansion coefficients about 6 times as great as mercury and freezing points below $-130°$ F.

In practice, the dry-bulb thermometer of a psychrometer is used to obtain current temperature readings.

Distant reading thermometers called telethermoscopes of the electric resistance type may also be used to obtain current temperatures.

MAXIMUM TEMPERATURE

Definition

Maximum temperature is the highest air temperature reached during a given period. It may be obtained by two methods: (1) by reading the highest temperature on the temperature-time

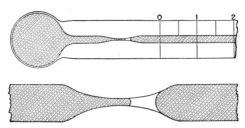

FIG. 50. Maximum thermometer bulb and stem constriction.

trace during a given period as recorded by a thermograph; or (2) by reading at the end of the period a maximum thermometer which has been set at the beginning of the period (Fig. 50).

Maximum Thermometer [11]

The instrument differs from an ordinary mercurial thermometer in having the bore of the tube constricted near the bulb so that mercury forced above the constriction by a rise in temperature after setting cannot readily return to the bulb. Maximum thermometers are filled with pure mercury at a temperature nearly equal to the highest temperature it can record. The space above the column at lower temperatures is therefore free of air. The maximum temperature for a given period after setting is found by carefully lowering the thermometer bulb to a vertical position and

Fig. 51. Townsend support for maximum and minimum thermometers.

reading the scale opposite the top of the mercury column. The thermometer is reset by whirling it until no further whirling will force the mercury farther into the bulb. It is then mounted with the bulb 5° above the horizontal. The thermometer is held firmly by a Townsend support or similar device which may be disengaged for whirling (Fig. 51).

If the thermometer is held in a horizontal position and then alternately tilted, first with the bulb higher than the stem and then with the bulb lower than the stem, it will be noticed that the thread of mercury may be made to flow to either end of the tube as desired.

Experience has shown that, in spite of every care in the inspection and testing of maximum thermometers, they will, if exposed vertically, sometimes fail to record the maximum temperature; that is to say, the constriction in the bore of the tube is not sufficiently fine to hold the mercury at the highest point reached, and, when the temperature falls after reaching the maximum point, the mercury in the column of the thermometer withdraws into the bulb and the record of the maximum temperature is lost.

When this occurs in the normal position, the thermometers are said "to retreat" and are called "retreaters."

A good maximum thermometer may be made a retreater by whirling it too violently.

Since it is known that a maximum thermometer may become a retreater without that fact being noticed, the safest procedure is to mount it with the bulb a little higher than the stem, as shown in Fig. 51. The support is so made that the angle of elevation of the thermometer back is 5°. In this position the column of mercury is not under pressure, and when the thermometer is lowered gently to reading position it will rarely fail to record the highest temperature reached.

When it is desired to make a reading, the thermometer must be slowly and carefully lowered to a position in which the mercury first comes to rest on the constriction. This reading position, or angle, is not the same for all maximum thermometers but must be determined by the observer watching the manner in which the mercury flows in the tube at different angles of elevation. Greater care must be used at high temperatures because of the relatively greater weight of the long column of mercury.

MINIMUM TEMPERATURE [11]

Definition

Minimum temperature is the lowest air temperature recorded during a fixed period. It may be obtained from thermograph readings or minimum thermometer readings. Minimum temperature of the air at the ground may also be obtained by placing the thermometer bulb at ground level.

Minimum Thermometers

Minimum thermometers provide a record of the lowest temperature occurring at the place of exposure from the time of setting until read. Uncolored ethyl alcohol is generally used to fill the tube, an operation which is done with the bulb immersed in a mixture having a temperature of nearly 0° F. Air is drawn into the tube at this relatively low temperature, and the tube is sealed. At higher temperatures the air exerts a pressure considerably in excess of that of the atmosphere. The excess largely prevents the diffusion of alcohol vapor into the air above the column and its subsequent condensation, thereby considerably increasing the accuracy of the thermometer. To provide against excessive internal air pressure at high temperatures, a small bulb is blown in the upper end of the tube. The bore of the tube is larger than that of a mercury thermometer to provide sufficient room for the index. This is a double-ended piece of dark-colored glass, so shaped as to follow the movement of the upper end of the alcohol column as the temperature falls, but to remain at the lowest point as the alcohol flows over it when the temperature rises. As shown in Fig. 51, when in the set position the thermometer is placed with

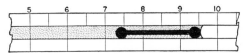

Fig. 52.   Minimum thermometer stem and index.

the bulb end about 5° below the horizontal. A scale reading at the end of the index farthest from the bulb is the lowest temperature at the place of exposure since the instrument was last set. The thermometer is set by turning it into a vertical position in the support, with bulb uppermost, when the index will fall to the end of the column and indicate the current temperature. The thermometer is then returned to the original position.

Usually the index will slide freely up and down the thermometer tube when one end of the tube is raised or lowered. Sometimes, however, and especially after the thermometer has been shipped, the alcohol column in the tube becomes broken up into short, detached segments and the index is frequently caught and held. To remedy this sometimes proves to be a difficult operation. Of course the thermometer cannot be used until the column is reunited. Observers are urged to learn and follow the instructions given below, and to notice carefully the several effects described so that when it becomes necessary to unite detached columns the danger of breaking the instrument may be avoided.

The fact that the vapor of alcohol condenses in the upper end of a minimum thermometer has always been a source of trouble to those who handle a considerable number of the thermometers. Manufacturers seal off minimum thermometers under considerable air pressure in order to lessen the probability of separation; but it is apparent that a marked fall of air temperature produces conditions favorable for condensation of the vapor of the alcohol, owing to a lessening of internal pressure, which in turn is due to the increased space above the column.

Under these circumstances, a good minimum thermometer may be misjudged as defective, when all that is needed to correct the condition is to reunite the column, and then to hang the instrument in a vertical position for an hour or more so that the alcohol adhering to the walls may drain down. The test for the success of the effort is simple: merely set the thermometer, immersed to the reading point, in melting, shaved, pure ice, or in melting snow. Within an hour or so it should read very close to 32°.

The many different ways in which an alcohol column may become separated make it impossible to unite it by any single method. Different processes are required not only for different conditions but also for different thermometers. Frequently, when there are only a few short, detached segments near the top, the index slides freely along the lower portion of the tube and drops into the bulb. Again, the detached segments may be found distributed along the tube with the index caught and held at some point above the main column. It is then advisable first to bring the index into the bulb as follows:

FIRST PROCESS. Hold the thermometer lightly between the thumb and fingers and strike the lower end of the metallic scale against the top of a padded table or other firm, padded object. One or two thicknesses of cloth or several folds of paper will provide a suitable pad to protect the thermometer from too severe a shock. The taps of the thermometer should be made lightly at first, and the index should be examined to see if it has moved along the tube even a slight distance. This can be told by noticing the exact position of the index with reference to the graduations on the tube or scale. If several taps fail to move the index, increase the force of the taps, a little at a time, until it begins to move, and then repeat the operation until the index penetrates the continuous column. Here it will fall of its own weight into the bulb. Usually this will be all that is necessary to place the index in the bulb. Sometimes, in the process, the detached column will have been partly or wholly united also. If the column is still broken in places, the observer should try a few more taps, and then examine it quickly and carefully. Small portions of the alcohol will generally be seen slowly moving along the sides of the tube toward the main column, and a continuation of the taps will unite the columns. Perhaps 15 or 20 minutes may be required to unite broken columns completely. If, however, the index cannot be made to move with sharp taps, or the columns cannot be united, it is advisable to try some of the methods described below, being careful always to avoid carrying any process so far as to endanger breaking the thermometer.

SECOND PROCESS. This process will not loosen the index but may unite the detached columns. Grasp the thermometer securely a little below the middle, with the bulb end down, and strike the edge of the metal back opposite the broken column sharply against the fleshy portion of the palm of the other hand, or, if necessary, against a small block of wood held in the hand. A continued jarring in this way often causes the alcohol to run down, though often a large number of taps is necessary. Observers should therefore not give up if the column does not unite at once but should watch very closely for the movements of small portions of alcohol along the sides of the tube. Here, again, care must be exercised not to strike too hard and also to hold the thermometer by the metal back in such a manner as not to squeeze, or press against, the stem of the thermometer itself. When the bore of the thermometer is large, this process is almost sure to unite the column. Good results are also obtained with thermometers of fine bore, although these, even in skilled hands, often require a half hour or more if the column is badly detached.

THIRD PROCESS. This process is sometimes effective in forcing the index into the bulb when other processes fail. Grasp the thermometer a little above the middle, clasping the fingers and hand firmly against the edges of the metallic back, but not so hard as to bring any pressure upon the glass tube, which should be turned toward you with the bulb uppermost. Holding the thermometer in this position at about the height of the head and with the arm free from the body, quickly lower the arm and hand a foot or more and turn the wrist at the same time so that the bulb of the thermometer describes a somewhat circular path downward through the air. Stop the motion with a sudden jerk just as the thermometer reaches the vertical. If the thermometer is grasped properly, a very violent motion can be given it in this way without danger of breaking it. It will sometimes be necessary to repeat the operation a considerable number of times to unite the detached columns entirely.

FOURTH PROCESS. A modification of the swinging process just described consists in whirling the thermometer rapidly on a short string. For this purpose a stout, doubled string is passed through the hole in the top of the metal back of the thermometer. The string is firmly grasped at a distance of 6 or 8 inches from the thermometer, which may then be given a very rapid whirling motion. Considerable care and practice are required to whirl the thermometer rapidly and stop it safely. This method will, however, often bring down the index and unite detached columns.

If observers are unsuccessful after carefully following the above instructions, the matter should be reported to the manufacturer with full particulars as to what has been done.

TESTING MINIMUM THERMOMETER. One or more of these processes should nearly always be sufficient to unite any detached column. When all the segments have been reunited and the index is properly placed, the thermometer may be checked for proper functioning and its operation noted as follows: Hold the thermometer vertically and warm the bulb by holding it in the hand; then turn the instrument upside down. Watch the index as it glides along the tube; when it strikes the top of the column it will come to a stop. The operation of bringing the index to the top of the column is called "setting" the thermometer. In practice, of course, it is done without warming the bulb.

Next hold the thermometer horizontally. As the bulb cools off, the index will be dragged backward toward the bulb by the top or end of the alcohol column. It is a good plan to hasten the cooling by placing a little wet cloth or piece of ice against the bulb. When the movement of the index ceases, warm the bulb again with the hand. The column will go up immediately, but the alcohol will flow around the index and leave it unmoved at the lowest point; that is, the index remains so that its *top* end is at the lowest point reached by the alcohol column, and the minimum temperature is indicated in this way. The thermometer must be held horizontally throughout these operations.

When the thermometer is not in use for observation, it is a good plan to hang it up, as bubbles are less likely to form in the tube in this position.

The thermometer should also be maintained in a vertical position for several hours after a broken column has been united to permit any alcohol clinging to the sides of the bore to drain down.

THERMOGRAPH

Types

Thermographs are of two types: those that are exposed and record in the instrument shelter, and those that are distant reading with a thermometer element exposed in the shelter. In the latter

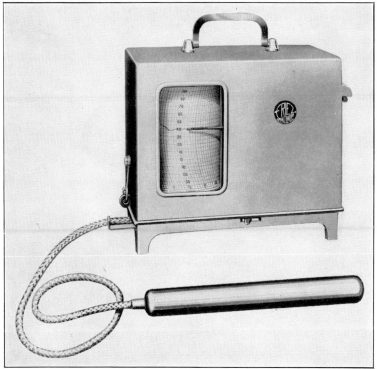

Courtesy Julien P. Friez and Sons, Inc.

FIG. 53. Remote-reading capillary-tube thermograph.

type, the record of temperature is transmitted to a recording device at a distance, usually in the observatory.

In most exposed thermographs the thermometer consists of a slightly curved metal alcohol-filled tube called a Bourdon tube. The curvature of the tube changes with changes in temperature. One end of the tube is rigidly fixed to the instrument case; the other end is connected by levers to a recording pen. A clock-rotated drum covered with recording paper receives the ink trace of temperature variations.

A second type has a bimetallic strip curved into a spiral and anchored at one end to the case. Variations in temperature cause the coil to wind or to unwind. The movement is communicated by the other end of the coil to a pen arm that bears on the recording paper.

Distant-reading thermographs are of two types. In one the thermometer bulb is about 12 inches long and 1 inch in diameter and contains mercury. A capillary tube connects the thermometer to a Bourdon tube on the recorder. The recorder is temperature compensated so that the true remote temperature only is recorded.

Another type of distant-reading thermograph uses a tube filled with vapor between the thermometer bulb and the recorder.

Care and Maintenance of Recording Device

There are only a few pointers on how to obtain good results with recording instruments.

1. Check the clock for accuracy. Make necessary adjustments to keep it running and to make it keep correct time. Always make a time mark on the sheet of recorder paper when placing it on the drum, and again when removing it. A sheet that runs for several days may require several intermediate time checks. Wind the clock carefully at each change period.

2. Place the sheet firmly around the drum and with the lower edge pressed tightly against the lip. Sheets are printed to read correctly only when this is done.

3. Clean the pen carefully and renew the ink at each change period. Some ink-and-paper combinations make it necessary to check the ink supply between paper changes.

4. To set the pen at the correct time, place the clock and drum on the shaft with the pen at the approximate time and then turn the cylinder forcibly, in a *counterclockwise* direction (as viewed from above), to take up all play until the pen reads correctly with respect to time.

The best time to check the temperature record is close to the time of minimum temperature, as the temperature is changing slowly and the differences of lag between the comparative thermometer and the element of the thermograph will be of least consequence.

Chapter VII

HUMIDITY

Definitions

Water exists in the atmosphere in three states: liquid, solid, and gaseous. The liquid form is rain, water clouds, and fog. The solid consists of snow, hail, and ice-crystal clouds. Humidity measurement is a method of expressing the amount of invisible or gaseous water vapor in the atmosphere.

Experiment has shown that there is an upper limit to the amount of water vapor that can be contained in a given space at a given temperature. Water vapor mixes with dry air and acts as an independent gas. We may speak of the pressure exerted by the dry air, and of the pressure exerted by the water vapor, or vapor pressure. When a given space contains the maximum amount of water vapor at a given temperature, the pressure exerted by the water vapor will also be the maximum and the space will be saturated. The saturated vapor pressure increases rapidly with increasing temperature. Before discussing methods for measuring humidity let us consider the various humidity definitions.

VAPOR PRESSURE (e) is the partial pressure exerted by the water vapor. The total pressure measured by a barometer is made up of the sum of the partial pressure of dry air and the vapor pressure. If the air is not saturated the vapor pressure (e) will be less than the saturation vapor pressure (e_s) for a given temperature. Vapor pressure may be expressed in millibars or other pressure units.

RELATIVE HUMIDITY (f) is defined as the ratio of the actual vapor pressure to the saturation vapor pressure at a given temperature. It is expressed in percentage in a scale from 0 per cent for air with no water vapor to 100 per cent for saturated air.

$$f = \frac{e}{e_s} \quad (e_s \text{ with respect to water at temperatures above and below } 32° \text{ F.})$$

ABSOLUTE HUMIDITY is the density of water vapor and is usually given in grains per cubic foot (7,000 grains = 1 lb.) or in grams per cubic meter. It is the amount of water vapor per unit volume.

DEW POINT (D.P.) is the temperature to which the air must be cooled at constant pressure to saturate it. Dew point is expressed in degrees Centigrade or Fahrenheit with respect to water.

SPECIFIC HUMIDITY (q) is the ratio of the density of the water vapor to the density of the mixture of dry air and water vapor. It is expressed in grams per kilogram of mixture.

A simple formula in which vapor pressure = e and total air pressure = p may be applied to obtain specific humidity q.

$$q = 622\,\frac{e}{p} \text{ grams per kilogram of moist air}$$

MIXING RATIO (w) is the ratio of the density of the water vapor to the density of dry air. A simple formula

$$w = 622\,\frac{e}{p - e} \text{ grams per kilogram of dry air}$$

where e is vapor pressure, p is total air pressure, and 622 is a constant.

WET-BULB TEMPERATURE (t_w) is the lowest temperature it is possible to produce by evaporating water or ice from the bulb of a thermometer. For low humidities the difference in temperature between a wet-bulb thermometer and a dry-bulb thermometer increases with increasing dry-bulb temperature. At saturation (100 per cent relative humidity) no evaporation will take place and both wet- and dry-bulb readings will be the same.

An equation may be written in terms of

e = vapor pressure in inches.
e' = pressure in inches of saturated vapor at t'.
t = temperature of the air in °F.
t' = temperature of wet-bulb thermometer in °F.
B = barometric pressure in inches.

$$e = e' - 0.000367B(t - t')\left(1 + \frac{t' - 32}{1571}\right)$$

THE OBSERVATION OF HUMIDITY

Obtaining atmospheric humidity measurements requires the same type of observation point as that for temperatures. The two instruments now in general use for measuring humidity, the hair hygrometer and the psychrometer, both require that the air should be circulated freely about them during the measurement. Both types of instruments are mounted in the instrument shelter.

PSYCHROMETERS [10, 11]

Ventilated wet- and dry-bulb psychrometers are used for relative-humidity observations and to determine the temperature of the dew point. Two types are common. One type is ventilated by whirling the thermometers, either by means of a whirling apparatus installed in a large shelter or by using a sling psychrometer outside a shelter. The second type is ventilated by drawing air past the psychrometer bulbs by means of a small hand-operated fan (Fig. 54). The psychrometer fan has come into general use, for it is adapted to both airway and cotton-region shelters. Two accurate mercury-in-glass thermometers are always required. Those used with the whirling apparatus are attached to counterbalanced arms, which are pinned to a spindle carrying a cast-iron pinion, meshed with a bevel gear, turned by an attached crankshaft. The whirling apparatus is securely screwed to the floor of a large instrument shelter with the crankshaft projecting through the front of the shelter. The proper ventilation of a psychrometer requires an air speed past the thermometers of about 15 feet per second.

The wet bulb of the psychrometer, and a short length of the stem above it, are covered with fine, loosely woven muslin, carefully washed to remove the sizing. Before a psychrometric reading is made, the muslin must be moistened with pure, clean water. The muslin must be replaced whenever it becomes dirty, to insure correct readings. To obtain accurate readings with a sling psychrometer, the instrument should be whirled in the shade, and the observer should face the wind to avoid temperature effects due to the heat of the body. When a psychrometer is used, the single exposed thermometer may be dispensed with.

When the relative humidity is low and the temperature is high, there will be a large depression of the wet-bulb temperature below that of the dry-bulb. At high relative humidity and low temperature the "depression" is small. It is therefore necessary to read both thermometers to tenths of a degree, especially when the depression is small.

Several other, more expensive types of psychrometers operate on the same principle as described above. The Assmann Psychrometer has an exhaust fan operated by an electric motor. The wet- and dry-bulb thermometers are mounted in an inlet duct

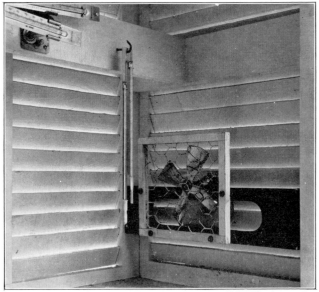

U.S.W.B. photograph.

Fig. 54. Wet- and dry-bulb psychrometer and fan.

through which air is drawn. In general, the velocity of the air past the wet bulb is more constant in the motor-driven fan-type psychrometer than in the hand-type, and readings are consequently more reliable.

Care of the Wet-bulb Thermometer [21]

Cleanliness

The muslin must be kept clean. If it is dirty, the readings will not be correct, since water will not evaporate at the same rate from dirty as from clean muslin.

1. Change the muslin at least once a week.
2. At coastal stations the muslin should be changed immediately after a storm with an onshore wind.
3. After a dust storm, the muslin should be changed immediately.

Wetting the Muslin

1. Use only soft water, distilled water, or rain or snow water.
2. Methods of moistening the muslin at different temperatures:

(a) Moderate temperatures above 32° F. (about 37°–70°): moisten just before taking a reading.

(b) Very high temperature and dry air: Wet the muslin thoroughly, submerging the wet bulb in the water long enough for it to reach equilibrium with the water temperature. Then wait a few minutes before whirling. Do not whirl too rapidly.

(c) Temperature close to, or below, freezing: (1) Moisten the muslin 10 to 15 minutes before taking the reading; if difficulty is found in obtaining a depression, the interval between moistening and reading may be increased to as much as 30 minutes. (2) Keep the water at room temperature, and, when wetting the muslin, make sure that any previous accumulation of ice on the bulb is melted off. An excessive accumulation of ice will act as insulation and prevent the thermometer from indicating the actual wet-bulb temperature.

PSYCHROMETRIC TABLES

Complete tables for reducing wet- and dry-bulb readings to relative humidity and dew point are for sale at the Government Printing Office, Washington 25, D. C.; the title is "Psychrometric Tables for Obtaining the Vapor Pressure, Relative Humidity, and Temperature of the Dew Point," Publication number W. B. 235.[20]

HYGROMETERS

The most common type of hygrometer is made of strands of human hairs. The characteristic of hair in elongating as the relative humidity increases is utilized as the basis of operation for this instrument. In general, the hair hygrometer is little used in meteorology except in the recording type of instrument. There is considerable lag in the response of the hair to changing values of relative humidity, and this lag increases as the temperature lowers. Rapid variations in relative humidity are impossible to record because of the slowness of the hairs' response.

Care of the Hair Hygrometer

The hairs should be kept clean by removing dust with a soft camel's-hair brush. They may be washed with a soft brush which

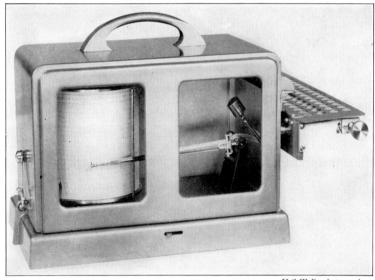

U.S.W.B. photograph.

Fig. 55.   Hair hygrograph.

has been dipped in distilled water. The hair strand should never be touched with the hand, as grease and oil prevent the hair from absorbing moisture. If the humidity is very low for a long period, it is well to expose the hygrometer to a saturated atmosphere about once a month.

The Recording Mechanism

The recording mechanism of the hygrograph operates on the same principle as that of the thermograph, and often the two instruments are combined to use the same clock drum and recording sheet. The care of pens and operation of the clock drum have been described in the chapter on temperature.

Chapter VIII

WIND

DEFINITIONS

Wind is defined as moving air, and the term is generally limited to air moving horizontally, or nearly so; vertical streams of air are usually called "currents." To describe wind at a given time, its direction and speed must be specified.

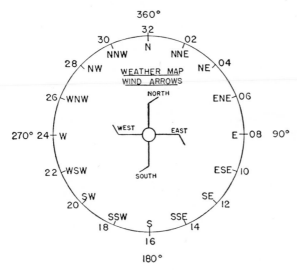

FIG. 56. Directions from point of observation.

WIND DIRECTION is that direction *from which* the wind is moving. Direction is expressed in points of the compass or degrees of azimuth as measured from true north (Fig. 56). If the wind is blowing from east toward the west the direction is regarded as *east* or 90°.

Wind direction is indicated by an anemoscope or wind vane which is designed to point into the wind. Thus the arrow points in the direction the wind is coming from. A wind vane pointing eastward indicates an east wind.

WIND SPEED is the speed of the air past a given point on the earth's surface. In the United States wind speed is expressed in units of miles per hour. In countries using the metric system speed may be expressed in kilometers per hour or in meters per second. For navigational purposes, especially by naval units, speed is expressed in knots (nautical miles per hour). Instruments for measuring wind velocity are called anemometers.

GUSTINESS. When the wind blows over a rough surface at moderate or high speed, small eddy currents arise and are carried along with the wind. The eddy currents range in size from a few feet to a hundred feet or more in diameter. If the wind speed is observed by means of a sensitive anemometer, it will be noticed that there will be periods of low speed and periods of high speed, with quite irregular variations between the two.

A gustiness factor or percentage gustiness is obtained by the following relation for a 10-minute period: [21]

Percent gustiness

$$= \frac{\text{Highest wind speed} - \text{Lowest wind speed}}{\text{Average wind speed}} \times 100$$

or

$$G\% = \frac{S_{maximum} - S_{minimum}}{S_{mean}} \times 100$$

EXPOSURE

The choice of a site for the exposure of the wind instruments is most important and may be regarded as of more importance than the choice of the wind instruments themselves.

The ideal anemometer exposure is secured when the instrument is placed where the movement of wind is unobstructed from any direction, such, for example, as would be obtained in the center of a large open expanse or plain. In a perfectly open or flat location the standard height of the wind vane and anemometer is 33 feet above the ground.[18] At most observatories only an approximation to the ideal exposure is possible. The location should be such, however, that the instruments are not sheltered by trees or buildings, for eddies or irregular vertical and horizontal air motions caused by such obstacles may extend both vertically and horizontally great distances and affect the accuracy of the wind data. In a city, the anemometer should be exposed above a high building in the vicinity of the observatory and as far as practicable above the immediate influence of the building itself.

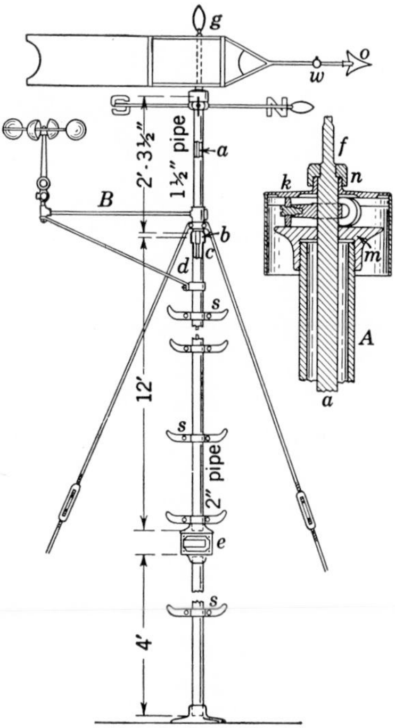

WIND-INSTRUMENT SUPPORT [10]

When an ideal exposure is not possible, wind instruments are usually supported by an iron pipe, 12 to 18 feet in length. The pipe should be fastened securely to a base plate fixed to a roof or to a platform and held in a vertical position by at least three guy rods (Fig. 58). To afford access to the top of the support, steps or foot rests should also be provided.

THE WIND VANE

The point of reference of the wind vane, true north, should be established by orientation methods described in Chapter II. The response of the vane to changes in wind direction

FIG. 57.   Anemometer and wind vane on support: *a* axis of vane (solid iron rod); *b* coupling; *c* joint; *d* indicates where one end of pipe is screwed firmly into the lower end of the solid rod of the vane; *e* contact box; *f* indicates where axis rod is squared; *g* spearhead nut; *k* carrier or frame for antifriction rollers; *m* cap; *n* shoulder formed on rod; *o* arrow point; *w* counterweight; *B* cross-arm support for anemometer; *s* foot rests; *A* support pipe and vane bearing detail.

depends principally upon the mass of the vane and the friction in the bearing. Most wind vanes have a large fin that offers resistance to the wind. The fin is attached to a rod that passes through a bearing, and it is counterbalanced by a heavy arrow point of small area. The vane is free to move about a vertical axis.

There are several methods for communicating the direction of the vane to a remote observing point. If the vane is located on

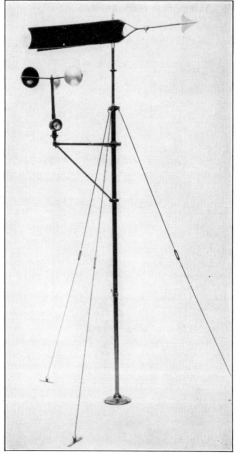

U.S.W.B. photograph.

Fig. 58.   Wind instrument support.

the roof of the observatory, the shaft may be extended down through the supporting pipe to an indicator in the room below. The pointer of the indicator may be geared to the vertical shaft so that the pointer will be in a vertical plane.

In general, electrical methods are found to be more suitable than mechanical methods. A commutator bearing or commutator arrangement operated by the movement of the vane may serve to close an electrical circuit connected with a series of electric bulbs in the observatory. Each bulb corresponds with a wind direction. There are 4-point and 8-point indicators. By means of the 8-point indicator, direction to 16 points of the compass can be obtained, as the commutator brush or commutator will light two bulbs when the direction falls between two of the principal points.

A second type of indicator employs two self-synchronous motors. One motor, actuated by vane movements, induces corresponding movements to the second, which is attached to a pointer in the observatory. This type is considered the best for direct reading and will give an accuracy of about 2°.

Recording Wind Direction

Automatic recording of wind direction by the United States Weather Bureau is accomplished by a *register*. A clock provides regular and uniform motion to a cylinder on which is placed a ruled sheet of paper adapted to receive the record. The record is traced by means of a pen attached to the armature of electromagnets.

Four electromagnets are connected to the cardinal direction contacts of the wind vane. A clock contact supplies current to the wind direction circuit once each minute. The circuit representing wind direction is closed, and a dot is made on the paper. Intermediate directions are indicated when two pens make dots at the same time, that is, if the N and E pens are depressed by their electromagnets, the wind direction is NE.

Anemometers

Four principal types of anemometers are used in meteorological work: rotating cups, pressure plates, bridled or torque type, and pressure tube anemometers.

Cup anemometers are in general use by the United States Weather Bureau. The Robinson cup anemometer, developed in 1846, has 3 or 4 cups extended about a vertical axis so that the plane of the cup is in a vertical position. By means of proper gear reductions the rotation of the cups is calibrated in terms of wind speed. One disadvantage of the cup anemometer is that in gusty winds the cups tend to accelerate quickly and lose speed slowly, thereby causing the average speed to be registered too high.

Wind speed may be indicated at a distance by a buzzer or by a light operated at $\frac{1}{60}$-mile or 1-mile intervals. The number of $\frac{1}{60}$-mile buzzes or flashes in 1 minute indicates the wind speed in

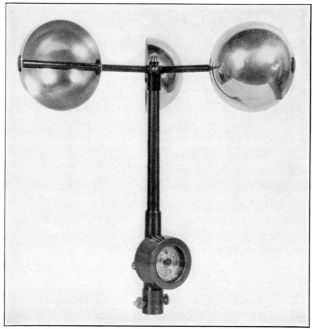

U.S.W.B. photograph.

Fig. 59.   Three-cup anemometer.

miles per hour. The speed may be recorded by a magnetically operated pen on a clock-driven drum. A 1- to 10-minute average wind seems to satisfy the requirements of meteorologists for use on weather maps. For climatological purposes a 5-minute average wind speed may be more desirable.

To observe and record gusty winds a more sensitive type of anemometer is required.

PRESSURE-PLATE OR PENDULUM ANEMOMETERS are perhaps the oldest type of wind-measuring instrument. The principle was first applied by Hooke in 1667.[3]

In order to obtain satisfactory results, the pendulum type should be calibrated by a standard anemometer. The period of swing of home-made instruments may be such that the pendulum will overswing or flutter and give incorrect wind readings.

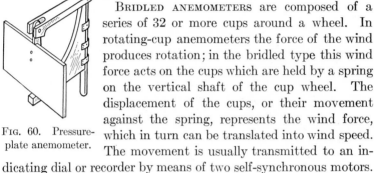

BRIDLED ANEMOMETERS are composed of a series of 32 or more cups around a wheel. In rotating-cup anemometers the force of the wind produces rotation; in the bridled type this wind force acts on the cups which are held by a spring on the vertical shaft of the cup wheel. The displacement of the cups, or their movement against the spring, represents the wind force, which in turn can be translated into wind speed. The movement is usually transmitted to an indicating dial or recorder by means of two self-synchronous motors.

FIG. 60. Pressure-plate anemometer.

THE PRESSURE-TUBE ANEMOMETER is the same instrument as that used on airplanes to measure air speed. The anemometer consists of a "pitot" tube which is kept pointed into the wind by a vane. The forward end of the tube is open to the wind pressure. It is necessary to measure the pressure difference between wind pressure and the "static" pressure or air pressure. A series of holes is arranged around an outer sleeve of the pitot head. Pressure difference is measured by a U-tube manometer or by a pressure gage. The design of the anemometer is based on the principle that the speed of the wind is proportional to the square root of the pressure difference divided by the air density. Since density changes with pressure and temperature variations, the speed obtained from an indicator based on *average density* may be in error. W. H. Dines first introduced the

Courtesy Julien P. Friez and Sons, Inc.

FIG. 61. Bridled anemometer

pressure-tube anemometer as a meteorological observing instrument in the year 1892.

In gusty winds the inflow and outflow of air from the tube and connecting pipes cause the fluctuations to be observed at slightly different times from the actual gusts; and the speed at the peak of the actual gust is therefore always greater than that recorded or observed on the indicator.[3]

The usual method of recording utilizes the Dines float manometer.

U.S.W.B. photograph.

Fig. 62.   Dines pressure-tube anemometer.

Gustiness

Studies in the structure of winds indicate that the sudden variations in the wind, usually designated as gustiness, are the manifestations of turbulence.

The turbulence that gives rise to "gusty" winds is of two distinct types. The first type, mechanical turbulence, is caused by irregularities in the terrain, which produce eddies in the wind. Thus, buildings and other obstructions in rough country give rise to gustiness, while over level country, or over the sea, mechanical turbulence is at the minimum. If the roughness of the surface in the neighborhood of a station varies with direction from the observatory, it follows that gustiness will vary with the wind direction.

The second type, thermal turbulence, is caused by the heating of air at the surface. When relatively cold air is blowing across heated ground, convection currents can develop easily, and thermal turbulence may become severe. If an inversion forms, however, the convection currents will be broken up and mechanical eddies will be damped, so that gustiness may be destroyed.

The eddies caused by thermal turbulence are generally larger in extent than those caused by mechanical turbulence. Though mechanical eddies are usually of approximately the same size as the obstructions that produce them, thermal eddies may extend laterally over a mile or more, and extend upward as convective currents for many thousands of feet.

Most states of gustiness may correctly be attributed to a combination of the two types of turbulence.

An anemogram of a gusty wind generally has the appearance of Fig. 63. Individual gusts have been analyzed (by means of special instruments), and it has been found that the speed increases rapidly to the peak and then falls more slowly to the lull (Fig. 63).[21] This trace shows the analysis of individual gusts as made by a Gurley instrument. The space between each two curved vertical lines represents ¾ minute.

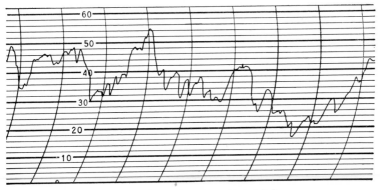

Fig. 63. Anemograph of gusty winds.

Other indications of gustiness may be found in the behavior of movable objects in the wind. Sudden more or less violent changes in the swaying of trees, etc., are common indications.

When instruments are not available and the gustiness has to be estimated by visual means, the specifications in the Beaufort scale should be used. For instance, if a wind makes sudden variations between "gentle" and "fresh," "fresh gusts" are indicated. If the wind makes sudden variations between "gentle" and "strong," we may be sure that there are "strong gusts."

A report of gustiness has definite value to an aircraft pilot in making landings, for the gustiness at his flying level may be entirely different from that near the surface. It is necessary to report gustiness especially for the less-experienced pilots. Sudden gusts of wind have actually overturned planes on landing.

Squalls [10]

The basic difference between gusts and squalls is in the time element. A gust takes only a few seconds or less to pass, whereas a squall usually lasts several minutes. Under certain conditions, thermal turbulence may develop into large systems, several square miles in area and many thousands of feet in height. Such an eddy may take a number of minutes to pass and may no longer be classified as a gust but must rather be considered a squall. Smaller variations due to mechanical turbulence are often superimposed upon thermal turbulence in a squall. Squalls are also associated with other meteorological factors such as clouds, precipitation or temperature changes.

Maximum Wind Velocity

For the purpose of synoptic weather reports, the "maximum wind velocity" until July 1939 was regarded by the United States Weather Bureau as the highest 5-minute mean wind speed and average direction. After July 1939, the "highest wind velocity" replaced the "maximum wind velocity"; however, this "highest wind velocity" was not for a 5-minute period but only for a 1-minute period. The new numeral code (1942) has reverted to the use of "maximum wind velocity," but it still refers to a 1-minute period.

The maximum wind may or may not be gusty. It will be noted that a *steady* maximum wind generally causes far less damage than a *gusty* maximum wind of the same average speed.

Estimating Wind Speed

All observers, at some time or other, are likely to encounter an anemometer that has become inoperative—a situation that necessitates the estimation of wind speeds. The quickest way to gain experience in estimating wind speeds is by conscious effort. When reading the wind speed, note the visible effects of the *wind pressure* at that particular speed. Try to memorize the general picture of flag movement, say, or the swaying of branches, twigs, grass, etc.

In a short time one can develop a keen sense of the relationship between wind force and wind speed, and indirect yet accurate estimations of the speeds that would be recorded by the observatory anemometer will be easy to make. Specifications for estimating the force of the wind on land are given in the Beaufort scale. The estimate of speed is obtained by noting the corresponding speed equivalents given in the table of wind-speed equivalents, Table VIII–4.

"Calm" is recorded only when the anemometer cups are *not moving* or when smoke rises *vertically* without noticeable drift.

HISTORICAL NOTE ON THE BEAUFORT SCALE *

Most observers (and some meteorologists) in the United States are unaware of the derivation of the wind speed equivalents of the Beaufort scale and of the specifications intended to govern their use. It therefore seems desirable to trace the history of our present scale and to remove, perhaps, several misconceptions.

Ever since man first began to use water-borne craft for transportation, the force of the wind has been of particularly great importance to him. So sailors learned to estimate the force of the wind in terms of a scale that specified the force as "calm," "air," "breeze," "gale," "storm," "hurricane." In addition, they qualified these terms by such adjectives as "light," "moderate," "fresh," and "strong."

These estimates of the force of the wind were based primarily upon the effects of the wind upon the ocean surface, effects with which they were thoroughly familiar. Seamen know, for instance, that the smallest ripples on the ocean surface respond markedly to sudden changes in wind; that an apparent darkening of the waters indicates the travel of individual gusts.

Further, their estimate of the wind force is based upon its *total* effect upon the surroundings. When a sailor estimates the wind

---

* This discussion was prepared by the United States Weather Bureau at the suggestion of Professor G. Emmons of New York University, and with the help of his notes, entitled "The Beaufort Scale of Wind Force and Its Velocity Equivalents." His original source, "The Velocity Equivalents of the Beaufort Scale," by G. C. Simpson (*London Meteorological Office Professional Notes* 44), was also used.

to be a moderate gale, it makes no difference to him, in making the estimation, whether he is standing on deck or is at the masthead. He would make the same estimate in either location. If an anemometer were placed at both points, however, it would probably be found that the one on deck would read, say, 30 mph, while the one at the masthead would read 35 mph. The point of observation affects the measure of speed but not the estimate of *force*.

In 1806, Admiral Sir Francis Beaufort devised a scale of wind force based upon the descriptive scale long used by sailors. He probably began by assigning a force number to each descriptive term, as in Table VIII-2.

Rather than depend upon these descriptive effects, Beaufort wanted to define his scale in terms of some standard object, in much the same way that a standard measure might be used to check the length of some other object. He therefore chose the typical British man-of-war of his time.

He had probably noted that, when a sailor described the wind as a light breeze, his "well-conditioned man-of-war, under all sail and clean full, would go in smooth water from 1 to 2 knots." In a similar manner, from his careful observations, he described the effect of each of the thirteen terms of wind force upon his man-of-war. The scale in Table VIII-3, except for a few slight verbal changes, is the same as that adopted by the British Navy in 1838.

The use of this scale spread from the Navy to the British mercantile marine, and eventually to ships of other nations.

Meanwhile, meteorology had begun to progress toward the use of synoptic reports, and in 1874 the International Meteorological Committee adopted the Beaufort scale for purposes of international telegraphic transmission of weather data.

As originally drafted, the Beaufort scale made no reference to the actual (velocity) speed of the wind. With the development of meteorology it became necessary to relate measured wind speeds to the Beaufort scale. It would have been simple to assign code numbers to an arbitrary scale of speeds and discard the Beaufort scale, but winds reported from ships must still be estimated to a large extent, and so a correlative scale was necessary. Even observers on land must from time to time make estimates of wind force.

For these reasons, many determinations of the speed equivalents were attempted, but results varied so widely as to be valueless. Finally, independent determinations by the London Meteorological Office and by the Deutsche Seewarte were made, under well-controlled conditions.

Experienced seamen stationed at exposed coastal or island stations where the sea was in full view made estimates of the wind force. At the same time the wind speed as indicated by the station anemometer was recorded. The estimated force was then correlated with the measured speed.

Now it was found that there was not a definite wind speed or even a definite interval of wind speeds for each Beaufort number. For instance, a wind estimated to be of force 7 by the British observer might have a speed of 38 mph as measured by the anemometer at his station. But a 38-mph wind at the German station might be estimated to be of force 9.

It is very improbable that two groups of experienced seamen, especially in view of the close contacts between seamen of all nationalities, would consistently show a difference of such a magnitude.

Upon study, it was found that the discrepancy was caused by the difference in the exposure of the instruments. The British exposures were usually free, as typified by the one at Scilly, where the anemometer was set 30 feet above the highest point of ground, itself 130 feet above sea level. The German anemometers, on the other hand, were mounted at comparatively low levels. We are therefore forced to conclude that the measured wind speed depends greatly upon the point of measurement (and also, is dependent upon the thermal properties of the air), whereas the estimate of wind force does not depend upon the point of observation. Thus, the exposure of the anemometer must be considered in converting wind speeds to Beaufort numbers.

The German meteorologist Hellmann found that wind speed varies with height in accordance with a logarithmic law.* This explains, in part, why the British and German investigations produced different results.

---

* Hellmann's logarithmic formula holds only when the lapse rate is adiabatic (stability is zero). Strictly speaking, therefore, a different table of Beaufort speed equivalents is required for a given anemometer exposure for each value of lapse rate. ("Lapse rate" is the rate of decrease in the temperature of the air per unit of height increase.)

The table of wind speed equivalents (Table VIII–4) used by the United States Weather Bureau gives the equivalents obtained in the British investigation in 1906. The values represent the average of the records of five anemometers, no two of which were similarly exposed.

It was determined that the values in this table would probably have been obtained from an exposure about 33 feet above level

TABLE VIII–1

*Standard Speed Equivalents of the Beaufort Scale as Prescribed by the International Meteorological Organization ***

Beaufort Number	Equivalent Limits of Speed, miles per hour
0	0– 1
1	2– 3
2	4– 7
3	8–11
4	12–16
5	17–21
6	22–27
7	28–33
8	34–40
9	41–48
10	49–56
11	57–65
12	over 65

* Not used by the United States Weather Bureau, however.

ground. It follows from the data of the Deutsche Seewarte investigation that anemometers exposed at much lower or higher levels would give different values from those in this table.

Since it was believed that, in general, the exposures at most stations could not be much better than those in the British investigation, and would probably not be much worse than those in the German investigation, the International Meteorological Organization, in 1926 or thereabouts, adopted a set of speed equivalents intermediate between those determined by the London Meteorological Office and those determined by the Deutsche Seewarte.

These equivalents (Table VIII–1) correspond with those predicted by Hellmann's formula for a height of about 20 feet above level ground and free from surrounding obstructions.

TABLE VIII-2

Specifications for Estimating the Force of the Wind at Sea

Beaufort Number	General Description of Wind	Criteria
0	Calm	Sea like a mirror.
1	Light air	Ripples with the appearance of scales are formed, but without foam crests.
2	Slight breeze	Small wavelets, still short but more pronounced; crests have a glassy appearance and do not break.
3	Gentle breeze	Large wavelets. Crests begin to break. Foam of glassy appearance. Perhaps scattered whitecaps.
4	Moderate breeze	Small waves, becoming longer; fairly frequent whitecaps.
5	Fresh breeze	Moderate waves, taking a more pronounced moderately long form; many whitecaps are formed. (Chance of some spray.)
6	Strong breeze	Large waves begin to form; the white foam crests are more extensive everywhere. (Probably some spray.)
7	Moderate gale	Sea heaps up and white foam from breaking waves begins to be blown in streaks along the direction of the wind.
8	Fresh gale	Moderately high waves of greater length; edges of crests begin to break into spindrift. The foam is blown in well-marked streaks along the direction of the wind.
9	Strong gale	High waves. Dense streaks of foam along the direction of the wind. Sea begins to "roll." Spray may affect visibility.
10	Whole gale	Very high waves with long, overhanging crests. The resulting foam, in great patches, is blown in dense white streaks along the direction of the wind. On the whole, the surface of the sea takes a white appearance. The rolling of the sea becomes heavy and shocklike. Visibility affected.
11	Storm	Exceptionally high waves. (Small and medium-sized ships might be for a time lost to view behind the waves.) The sea is completely covered with the long, white patches of foam lying along the direction of the wind. Everywhere the edges of the wave crests are blown into froth. Visibility affected.
12	Hurricane	The air is filled with foam and spray. Sea completely white with driving spray; visibility very seriously affected.

133

According to the further recommendations of the International Meteorological Organization, in cases where the exposure of the anemometer differs from this standard in such a manner that the effect is known, a special table of speed equivalents should be constructed applicable to the particular anemometer. Thus, if an anemometer is exposed at a height of 66 feet above ground, free from obstructions, the conversion to Beaufort numbers would be made by means of a table in which each speed equivalent in Table VIII–1 is multiplied by 1.23. This would give the speed that would have been recorded if the anemometer were at the standard level of 33 feet instead of at a level of 66 feet.

TABLE VIII–3

The Original Beaufort Scale

Beaufort Number	Beaufort's Description of the Wind	Beaufort's Criteria	
0	Calm		
1	Light air	Just sufficient to give steerage way.	
2	Light breeze	with which a well-conditioned man-of-war, under all sail, and clean full, would go in smooth water from	1 to 2 knots.
3	Gentle breeze		3 to 4 knots.
4	Moderate breeze		5 to 6 knots.
5	Fresh breeze	in which the same ship could just carry close hauled	Royals, etc.
6	Strong breeze		Single-reefs and topgallant sails.
7	Moderate gale		Double-reefs, jib, etc.
8	Fresh gale		Triple-reefs, courses, etc.
9	Strong gale		Close-reefs, and courses.
10	Whole gale	with which she could only bear close-reefed main topsail and reefed foresail.	
11	Storm	with which she could be reduced to storm staysails.	
12	Hurricane	to which she could show no canvas.	

Up to the present, however, this is not done in the United States Weather Bureau, probably because, in spite of the studies already made concerning surface winds, it is felt that not enough is yet known to warrant the application of the above exposure corrections. Not only would it be necessary to make a study of each station exposure, but, to be strictly accurate, it would also be necessary to make several tables incorporating a correction for different values of lapse rate at each observing station. This would be a tremendous undertaking.

The use of the Beaufort scale is not confined to coding wind speeds. At times an observer may need to report wind speed without the aid of an anemometer. At such a time he may estimate the force of the wind according to the descriptive terms in Table VIII–4. The speeds indicated, it should be remembered, are for a level about 33 feet above level ground. It should also be

TABLE VIII–4

Wind-Speed Equivalents

Descriptive Word	Velocity, miles per hour	Specifications for Estimating Velocities
Calm	Less than 1	Smoke rises vertically.
	1 to 3	Direction of wind shown by smoke drift but not by wind vanes.
Light	4 to 7	Wind felt on face; leaves rustle; ordinary vane moved by wind.
Gentle	8 to 12	Leaves and small twigs in constant motion; wind extends light flag.
Moderate	13 to 18	Raises dust and loose paper; small branches are moved.
Fresh	19 to 24	Small trees in leaf begin sway; crested wavelets form on inland waters.
	25 to 31	Large branches in motion; whistling heard in telegraph wires; umbrellas used with difficulty.
Strong	32 to 38	Whole trees in motion; inconvenience felt in walking against the wind.
	39 to 46	Breaks twigs off trees; generally impedes progress.
Gale	47 to 54	Slight structural damage occurs (chimney pots and slate removed).
	55 to 63	Trees uprooted; considerable structural damage occurs.
Whole gale	64 to 75	Rarely experienced; accompanied by widespread damage.
Hurricane	Above 75	

remembered that the Beaufort scale does not incorporate a correction for turbulence, or gustiness. If the observer does not take gustiness into account when it exists, then, in using the movement of trees, etc., as a measure of wind force, it is possible that his estimated value of force, and hence of speed, will be too high.

Chapter IX

PRESSURE [8]

GENERAL

In meteorology, pressure represents the force per unit area exerted by the atmosphere. At any given point it represents the total weight of a column of air of unit area extending to the outer limits of the atmosphere. The measurement of pressure at the ground or in the free air is affected by wind relative to the barometer or its housing. The effect of wind upon pressure is called the dynamic (motion) effect. In gusty winds the dynamic effect produces deviations from the true static (at-rest) pressure. Pressure is measured by various types of barometers, such as liquid or mercury barometer, and the elastic or aneroid (without liquid) barometer.

SELECTING THE OBSERVATION POINT [10]

The location of the barometer should be in a room of fairly even temperature. The direct rays of the sun should not fall on the instrument. Changes in temperature in different parts of the instrument cause unequal expansion of the scale and mercury, and readings will be in error. If the building is air conditioned, pressure inside may not be the same as that outside, and it will be necessary to provide the barometer with an airtight housing connected by a metal tube to the outside air. The outside opening of the metal tube, or static tube, should be arranged so that changes of wind direction will not introduce dynamic effects.

The simplest method is to use a static-head tube mounted on a wind vane. Heat for de-icing the tube in cold climates should be provided, and a water trap should be installed to prevent passage of liquid water into the barometer. (See Fig. 62.)

Stations located on or very near a mountain may have difficulty in measuring the true static pressure, because the blocking effect of the mountain may cause large variations in pressure.

Pressure Units

The unit of pressure is based upon a force unit per area unit. In the English system the pound per square inch, and in the metric system the dyne per square centimeter, are the basic units. The dyne is a unit of force. When 1 dyne of force is applied to a mass of 1 gram it will be accelerated 1 centimeter per second per second. In meteorology, the millibar, which is equal to 1,000 dynes per square centimeter, is the most common unit. The height of a column of mercury when reduced to standard gravity and temperature is also commonly employed.

The observed vertical length of a column of mercury is corrected in such a manner as to obtain the height that would exist if the temperature of the mercury were 32° F. (0° C.) and the column were subjected to a standard acceleration of gravity. The pressure produced by a column of liquid mercury is equal to the product of the density of the mercury, the height of the column, and the acceleration of gravity. The density of pure mercury is 13.5951 grams per cubic centimeter at a temperature of 32° F. at normal pressure (76 centimeters of mercury); and the acceleration of gravity is 980.665 centimeters per second per second at latitude 45° 29′. The following table presents a comparison of various units.

Inches of Mercury	Millimeters of Mercury	Millibars	Pounds per Square Inch
29.9212	760	1,013.25	14.7
1.00	25.4	33.86	0.49
0.0394	*1.00*	1.33	0.0193
0.0296	0.75	*1.00*	0.0145
2.036	51.6	68.9	*1.00*

Mercurial Barometers [21]

The most accurate practical method of measuring atmospheric pressure is by means of the mercurial barometer, which simply involves the accurate measurement of the length of a vertical column of mercury. There are various types of mercurial barometers. A common one is the Fortin type (Fig. 64).

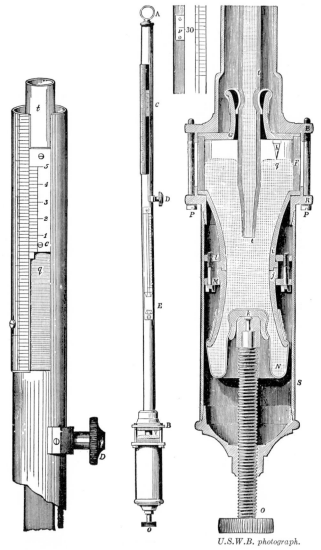

U.S.W.B. photograph.

FIG. 64. Mercurial barometer with Fortin cistern: *B* collar; *C* vernier; *D* milled head; *E* attached thermometer; *F* short glass cylinder; *G* boxwood cap; *M* split ring clamp; *N* kid leather bag; *O* adjusting screw; *P* screws; *R* flange; *S* portion of cistern; *h* ivory point; *i* and *j* curved boxwood pieces; *l* split ring clamp; *q* mercury; *t* glass tube.

Verticality of Mercurial Barometers

Some barometers swing slightly free within the ring support around the cistern, if the ring support is not fitted with clamping screws.

When such a barometer is used, the observer must be careful to adjust the barometer so that it is steadied against the ring. Figure 65 illustrates the correct position of the ivory point with respect to the point at which the cistern is supported against the ring.

Thus adjusted, the barometer can tilt slightly toward *B*. This, however, will only slightly affect the position of the mercury level in the cistern with respect to the ivory point.

However, if the cistern is rested against *C* or *D*, the barometer may tilt toward *D* or *C*, respectively. Then the position of the mercury level will be sensibly changed with reference to the ivory point as shown by Fig. 66.

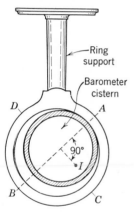

Fig. 65. Correct position of barometer cistern with respect to ring support (*I*, ivory point).

Use of the Vernier

The vernier is named after its inventor, Paul Vernier. It is a device for accurately measuring much smaller fractional subdivisions of a graduated scale than could be observed by the naked eye. For instance, a barometer vernier having ten subdivisions per inch measures to hundredths of an inch and allows an estimate to thousandths. One with 20 subdivisions to the inch measures to 0.002 inch.

On the 10-division vernier, 9 divisions on the barometer scale are divided into ten spaces on the vernier. Thus, each space on the vernier is $\frac{1}{10}$ part shorter than a space on the scale. Since the scale represents inches and tenths, the difference between the length of a space on the vernier and one on the scale is $\frac{1}{10}$ of $\frac{1}{10}$ of an inch, or $\frac{1}{100}$ of an inch.

On the 20-division-to-the-inch vernier, 24 parts on the scale are divided into 25 parts on the vernier. In this case $\frac{1}{25}$ of $\frac{1}{20}$ of an inch equals $\frac{1}{500}$ or 0.002 of an inch.

READING THE VERNIER. Since the spaces on the vernier are one-tenth smaller than those on the scale, then, in the adjustment drawn in Fig. 67*a*, the first line above the zero on the vernier is one-tenth part of the space, the next line two-tenths, the next, three-tenths, etc., distant from the line next above on the scale.

When, therefore, we find the vernier in such a position as shown in Fig. 67*b*, where the *fifth* line on the vernier coincides with a

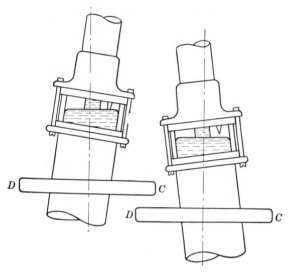

FIG. 66.   Effect of tilt in wrong direction on mercury level in the cistern.

scale line, it is clear that the zero line of the vernier must be just *five* tenths above the scale line next below it. Now, since the scale represents inches and tenths, Fig. 67*b* will read exactly 30.150 inches.

Often, no single line on the vernier will exactly coincide with a scale line, but *two* lines will be *nearly* in coincidence with two scale lines. One vernier line will be a little above a scale line while the next vernier line will be a little below a corresponding scale line.

In Fig. 67*c*, the seventh and eighth vernier lines are each nearly in coincidence with scale lines, but neither is *exactly* so. This indicates that the reading is somewhere between 30.27 and 30.28. Moreover, we can see clearly that the eighth line is nearer coincidence than the seventh. Therefore we estimate that the true reading is nearer 30.28, and call it 30.277. We might just as well

have called it 30.278. Either would be as nearly correct as
possible.

We can see, from the above, that readings can and should be
accurate within 0.002 of an inch.

In estimating fractions, the eye should be exactly in front of
the lines being studied.

Reading the 20-division-to-the-inch vernier: (1) This vernier
has 25 divisions. The 5th, 10th, 15th, 20th, 25th divisions repre-

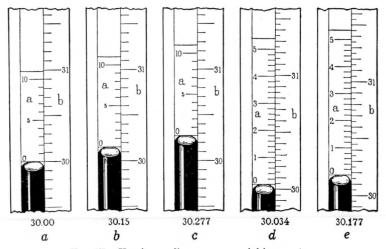

30.00   30.15   30.277   30.034   30.177
  a   *b*   *c*   *d*   *e*

FIG. 67.   Vernier readings on mercurial barometers.

sent 1, 2, 3, 4, and 5 hundredths of an inch, and are so marked.
Each single division represents 0.002 of an inch.   (2) In the use
of this vernier, if a reading, such as shown in Fig. 67*e*, is between
30.176 and 30.178, we do not try to estimate which is nearer, but
simply take *halfway*, or 30.177.   (3) *Caution!* Whenever the zero
line of this type of vernier is *above* one of the shortest lines of the
scale, be sure to take into account the 0.050 represented by the
short line.

Thus, in Fig. 67*e*, the coincidence on the vernier is between
lines designated 26 and 28, which corresponds to a reading between
0.026 and 0.028, or 0.027.   On the scale, however, the reading is
30.150 (0.050 of this is represented by the short line).   Thus the
total *correct* reading is 30.150 plus 0.027, or 30.177.

If the 0.050 is ignored, an *incorrect* reading of 30.127 would be
obtained.

Reading the Mercurial Barometer

1. Read the attached thermometer to the nearest one-half degree. Record the readings as 71.0°, 72.5°, etc.

2. Slowly turn the milled-head adjusting screw, which extends from the barometer at the bottom of the cistern, until the surface of the mercury in the cistern just touches the ivory point. In a newly cleaned barometer this contact can be observed by noting the meeting of the ivory point with its reflected image, or by a slight dimple in the mercury surface. If the mercury surface has become oxidized, the correct position can best be determined by sighting across the mercury surface against a frosted white glass pane that is sometimes built into the barometer; also, for these operations, an electric hand lamp or flashlight is nearly indispensable.

3. Tap the barometer case smartly with the finger in the neighborhood of the meniscus. Tapping improves the evenness of distribution of the mercury at the top of the column.

4. Turn the vernier adjustment by means of the milled-head screw projecting from the right of the barometer until the line of sight across the front and rear sighting edges on the lower portion of the vernier scale *just cuts off light from across the* **extreme summit** *of* the meniscus. A white background is essential. White paper serves well if renewed occasionally.

5. Read the height of the column. Inches and tenths are taken from the barometer scale. Hundredths are read on the vernier scale by selecting on the vernier the graduation that most nearly coincides with or matches a line on the main scale. When readings to thousandths of an inch are made, the third decimal figure is estimated from the vernier.

Determination of Actual Pressure from the Barometer Reading

Corrections must be applied to the barometer reading to determine the actual pressure.

1. "Constant" corrections. Certain errors, peculiar to a particular barometer and to a given geographical location of a barometer, are constant for the particular barometer at a particular place. These corrections comprise

(a) Instrumental correction
(b) Latitude correction ⎫ Gravity correction
(c) Altitude correction ⎭

These corrections are called the "sum of corrections" and are issued (by the United States Weather Bureau) on correction cards (W. B. Form 1059–Met'l) for each station and instrument. This card or its equivalent should always be posted near the barometer, preferably in the case or on the board directly behind the instrument.

2. Temperature correction. This correction depends upon the temperature of the mercury and the metallic scale. Tables are available giving the corrections for temperature measured by the attached thermometer.

3. Some stations have "constant" corrections and temperature corrections combined on a special table for easier computations. These corrections are then the same as the "total correction" described in the following paragraph and are similarly applied to the barometer reading in computing the "station pressure."

Application of Corrections

1. Add the temperature correction to the "Sum of Corrections" algebraically, to give the "total correction"; that is: (*a*) if their signs are *different*, *subtract* them and prefix the sign of the larger correction to the remainder; (*b*) if their signs are the *same*, *add* them and prefix their common sign to the sum.

2. Add the total correction algebraically to the observed barometer reading.

EXAMPLE.

Attached thermometer reading	76.5°
Observed barometer reading	30.287
Temperature correction	−0.131
Sum of corrections (from form 1059–Met'l)	+0.032
Total correction	−0.099
	30.287
	−0.099
Corrected pressure, or station pressure	30.188

Definitions of barometer heights are given in Chapter II.

Since the attached thermometer is read to the nearest one-half degree, only a single interpolation for pressure reading is necessary. For example: observed pressure reading = 30.287; attached ther-

Pressure

TABLE IX–1

Correction to Mercurial Barometer for Temperature of Attached Thermometer

°F.	24	24.5	25	25.5	26	26.5	27	27.5	28	28.5	29	29.5	30	30.5	31
							Observed reading of the barometer, in inches								
							SUBTRACT								
67	0.083	0.085	0.087	0.089	0.090	0.092	0.094	0.095	0.097	0.099	0.101	0.102	0.104	0.106	0.108
67.5	.084	.086	.088	.090	.092	.093	.095	.097	.098	.100	.102	.104	.106	.107	.109
68	.085	.087	.089	.091	.093	.094	.096	.098	.100	.102	.103	.105	.107	.109	.110
68.5	.087	.088	.090	.092	.094	.096	.097	.099	.101	.103	.105	.106	.108	.110	.112
69	.088	.089	.091	.093	.095	.097	.099	.100	.102	.104	.106	.108	.110	.111	.113
69.5	.089	.091	.092	.094	.096	.098	.100	.102	.104	.105	.107	.109	.111	.113	.115
70	.090	.092	.094	.095	.097	.099	.101	.103	.105	.107	.109	.110	.112	.114	.116
70.5	.091	.093	.095	.097	.098	.100	.102	.104	.106	.108	.110	.112	.114	.116	.117
71	.092	.094	.096	.098	.100	.102	.103	.105	.107	.109	.111	.113	.115	.117	.119
71.5	.093	.095	.097	.099	.101	.103	.105	.107	.109	.110	.112	.114	.116	.118	.120
72	.094	.096	.098	.100	.102	.104	.106	.108	.110	.112	.114	.116	.118	.120	.122
72.5	.095	.097	.099	.101	.103	.105	.107	.109	.111	.113	.115	.117	.119	.121	.123
73	.096	.098	.100	.102	.104	.106	.108	.110	.112	.114	.116	.118	.120	.122	.124
73.5	.097	.099	.101	.103	.105	.108	.110	.112	.114	.116	.118	.120	.122	.124	.126
74	.098	.101	.103	.105	.107	.109	.111	.113	.115	.117	.119	.121	.123	.125	.127
74.5	.100	.102	.104	.106	.108	.110	.112	.114	.116	.118	.120	.122	.124	.126	.129
75	.101	.103	.105	.107	.109	.111	.113	.115	.117	.119	.122	.124	.126	.128	.130
75.5	.102	.104	.106	.108	.110	.112	.114	.117	.119	.121	.123	.125	.127	.129	.131
76	.103	.105	.107	.109	.111	.113	.116	.118	.120	.122	.124	.126	.128	.131	.133
76.5	.104	.106	.108	.111	.113	.115	.117	.119	.121	.123	.125	.128	.130	.132	.134
77	.105	.107	.109	.112	.114	.116	.118	.120	.122	.125	.127	.129	.131	.133	.136
77.5	.106	.108	.110	.113	.115	.117	.119	.121	.124	.126	.128	.130	.133	.135	.137
78	.107	.109	.112	.114	.116	.118	.120	.123	.125	.127	.129	.132	.134	.136	.138
78.5	.108	.110	.113	.115	.117	.119	.122	.124	.126	.128	.131	.133	.135	.137	.140
79	.109	.112	.114	.116	.118	.121	.123	.125	.127	.130	.132	.134	.137	.139	.141
79.5	.110	.113	.115	.117	.120	.122	.124	.126	.129	.131	.133	.136	.138	.140	.143
80	.111	.114	.116	.118	.121	.123	.125	.128	.130	.132	.135	.137	.139	.142	.144
80.5	.112	.115	.117	.120	.122	.124	.127	.129	.131	.134	.136	.138	.141	.143	.145
81	.114	.116	.118	.121	.123	.125	.128	.130	.132	.135	.137	.140	.142	.144	.147
81.5	.115	.117	.119	.122	.124	.127	.129	.131	.134	.136	.139	.141	.143	.146	.148
82	.116	.118	.121	.123	.125	.128	.130	.133	.135	.137	.140	.142	.145	.147	.149
82.5	.117	.119	.122	.124	.127	.129	.131	.134	.136	.139	.141	.144	.146	.148	.151
83	.118	.120	.123	.125	.128	.130	.133	.135	.138	.140	.142	.145	.147	.150	.152
83.5	.119	.121	.124	.126	.129	.131	.134	.136	.139	.141	.144	.146	.149	.151	.154
84	.120	.123	.125	.128	.130	.133	.135	.138	.140	.143	.145	.148	.150	.153	.155
84.5	.121	.124	.126	.129	.131	.134	.136	.139	.141	.144	.146	.149	.151	.154	.156
85	.122	.125	.127	.130	.132	.135	.137	.140	.143	.145	.148	.150	.153	.155	.158
85.5	.123	.126	.128	.131	.134	.136	.139	.141	.144	.146	.149	.152	.154	.157	.159
86	.124	.127	.130	.132	.135	.137	.140	.142	.145	.148	.150	.153	.155	.158	.161
86.5	.125	.128	.131	.133	.136	.138	.141	.144	.146	.149	.152	.154	.157	.159	.162
87	.126	.129	.132	.134	.137	.140	.142	.145	.148	.150	.153	.155	.158	.161	.163
87.5	.128	.130	.133	.136	.138	.141	.144	.146	.149	.151	.154	.157	.159	.162	.165
88	.129	.131	.134	.137	.139	.142	.145	.147	.150	.153	.155	.158	.161	.163	.166
88.5	.130	.132	.135	.138	.141	.143	.146	.149	.151	.154	.157	.159	.162	.165	.168
89	.131	.134	.136	.139	.142	.144	.147	.150	.153	.155	.158	.161	.164	.166	.169

TABLE IX-2

Rock Springs, Wyoming—Airport—Station Elevation 6,745 Feet

Reduction of Barometric Pressure to Sea Level

(Tabular values are sea-level pressure)

Mean Temperature, °F.	Station Pressure, inches										
	23.70	23.80	23.90	24.00	24.10	24.20	24.30	24.40	24.50	24.60	24.70
30	30.49	30.62	30.75	30.88	31.00	31.13	31.26	31.39	31.52	31.65	31.78
32	30.47	30.59	30.72	30.85	30.98	31.11	31.24	31.37	31.50	31.63	31.75
34	30.44	30.57	30.70	30.83	30.96	31.09	31.22	31.34	31.47	31.60	31.73
36	30.42	30.55	30.68	30.81	30.93	31.06	31.19	31.32	31.45	31.58	31.71
38	30.40	30.53	30.65	30.78	30.91	31.04	31.17	31.30	31.42	31.55	31.68
40	30.37	30.50	30.63	30.76	30.89	31.02	31.14	31.27	31.40	31.53	31.66
42	30.36	30.48	30.61	30.74	30.87	31.00	31.12	31.25	31.38	31.51	31.64
44	30.34	30.46	30.59	30.72	30.85	30.98	31.10	31.23	31.36	31.49	31.61
46	30.32	30.44	30.57	30.70	30.83	30.96	31.08	31.21	31.34	31.47	31.59
48	30.30	30.43	30.55	30.68	30.81	30.94	31.06	31.19	31.32	31.45	31.57
50	30.28	30.41	30.53	30.66	30.79	30.92	31.04	31.17	31.30	31.43	31.55
52	30.26	30.39	30.52	30.65	30.77	30.90	31.03	31.15	31.28	31.41	31.54
54	30.25	30.37	30.50	30.63	30.76	30.88	31.01	31.14	31.27	31.39	31.52
56	30.23	30.36	30.49	30.61	30.74	30.87	30.99	31.12	31.25	31.38	31.50
58	30.22	30.34	30.47	30.60	30.72	30.85	30.98	31.11	31.23	31.36	31.49
60	30.20	30.33	30.45	30.58	30.71	30.83	30.96	31.09	31.22	31.34	31.47
62	30.19	30.31	30.44	30.57	30.69	30.82	30.95	31.07	31.20	31.33	31.45
64	30.17	30.30	30.43	30.55	30.68	30.81	30.93	31.06	31.18	31.31	31.44
66	30.16	30.28	30.41	30.54	30.66	30.79	30.92	31.04	31.17	31.30	31.42
68	30.14	30.27	30.40	30.52	30.65	30.78	30.90	31.03	31.15	31.28	31.41
70	30.13	30.26	30.38	30.51	30.63	30.76	30.89	31.01	31.14	31.27	31.39
72	30.12	30.24	30.37	30.49	30.62	30.75	30.87	31.00	31.12	31.25	31.38
74	30.10	30.23	30.35	30.48	30.61	30.73	30.86	30.98	31.11	31.24	31.36
76	30.09	30.21	30.34	30.47	30.59	30.72	30.84	30.97	31.10	31.22	31.35
78	30.07	30.20	30.33	30.45	30.58	30.70	30.83	30.95	31.08	31.21	31.33
80	30.06	30.19	30.31	30.44	30.56	30.69	30.81	30.94	31.07	31.19	31.32
82	30.05	30.17	30.30	30.42	30.55	30.68	30.80	30.93	31.05	31.18	31.30
84	30.03	30.16	30.28	30.41	30.54	30.66	30.79	30.91	31.04	31.16	31.29
86	30.02	30.15	30.27	30.40	30.52	30.65	30.77	30.90	31.02	31.15	31.27
88	30.01	30.13	30.26	30.38	30.51	30.63	30.76	30.88	31.01	31.13	31.26
90	29.99	30.12	30.24	30.37	30.49	30.62	30.75	30.87	31.00	31.12	31.25

mometer reading = 76.5°. Entering Table IX-1 for 76.5°, we find that the correction for 30.287 must lie about midway between that for 30.0 and 30.5, or midway between −0.130 and −0.132, which gives us a value of *−0.131.*

REDUCTION OF PRESSURE TO SEA LEVEL *

Since the altitudes of stations throughout the country vary widely, their station pressures cannot be easily compared. To make them all comparable, a reduction to sea level is made of each station pressure to determine what the pressure *would be if the station were at sea level.* Thus, the pressures at all stations are referred to the same elevation—standard sea level.

Pressure reductions have been computed for each United States Weather Bureau station and put in the form of tables, so that the observer's task is simplified. (A graphical method of pressure reduction is given in the Appendix.)

1. A separate table is available for each elevation. For low elevations (up to 50 feet above mean sea level) a constant correction to sea level is applied and no sea level reduction tables are used.

2. The table for a given high elevation concerns two quantities: (*a*) the station pressure; (*b*) the mean 12-hour *outdoor* temperature.

Method for Determining Mean 12-Hour Temperature

1. At stations making observations at least every 12 hours:

Mean 12-hour temperature

$$= \frac{\text{Current temperature} + \text{Temperature 12 hours ago}}{2}$$

2. At stations making only one observation a day, temperature 12 hours ago is obtained from the (corrected) thermograph trace. When the thermograph trace is used to obtain the mean temperature, take the temperature only to the nearest whole degree.

EXAMPLE 1.

Current temperature	52.6°
Temperature 12 hours ago	38.2°
Sum	90.8°
Mean = Sum divided by 2	45.4°

* The Report of the Chief of the Weather Bureau for 1900–1901 contains a description of the standard system of pressure reduction for the United States.

EXAMPLE 2.

Current temperature	60.8°
(Thermograph) temperature 12 hours ago	48.0°
Sum	108.8°
Mean = Sum divided by 2	54.4°

Use of Barometer Reduction Tables

1. The station pressure and the mean temperature value, used in the tables to find the sea-level pressure, are called the pressure argument and the temperature argument, respectively.

2. The pressure argument is the station pressure to hundredths of an inch only.

The third decimal is disposed of according to the following United States Weather Bureau *rule for disposal of decimals.*

(*a*) If the decimal figure to be disposed of is greater than 5 (or 5 with a remainder) the preceding figure will be increased by 1.

EXAMPLES. 29.806 becomes 29.81; 29.816 becomes 29.82; 29.826 becomes 29.83.

(*b*) If the decimal figure to be disposed of is 5 exactly, the preceding figure, when odd, will be increased by 1, and when even, it will remain unchanged.

EXAMPLES. 29.805 becomes 29.80; 29.815 becomes 29.82; 29.825 becomes 29.82.

(*c*) If the decimal figure to be disposed of is less than 5, the preceding figure will be retained unchanged.

EXAMPLES. 29.804 becomes 29.80; 29.814 becomes 29.81; 29.824 becomes 29.82.

3. It will be necessary to interpolate for hundredths when the station pressure falls between the tenths of an inch as given at the top of Table IX–2. Use Table IX–3 for interpolation.

4. The temperature value to use in the sea-level reduction table (Table IX–2) is the one nearest the mean temperature.

If the mean temperature is *exactly* halfway between two temperatures in the table, use the lower temperature in the table.

EXAMPLES. 1. Mean temperature = 58.7°. Temperature to be used in Table IX–2 = 58°.

2. Mean temperature = 59.2°, table temperature = 60°.

3. Mean temperature = 59.0°, table temperature = 58°.

Note: The above rule is arbitrary, to eliminate interpolation for temperature. The errors resulting by this approximation are *small*.

In all the following examples, use Tables IX–2 and IX–3.

TABLE IX–3

HUNDREDTHS FIGURE OF STATION PRESSURE	SEA-LEVEL PRESSURE DIFFERENCE (For one value of mean temperature)				
	0.10	0.11	0.12	0.13	0.14
1	0.01	0.01	0.01	0.01	0.01
2	.02	.02	.02	.03	.03
3	.03	.03	.04	.04	.04
4	.04	.04	.05	.05	.06
5	.05	.06	.06	.06	.07
6	.06	.07	.07	.08	.08
7	.07	.08	.08	.09	.10
8	.08	.09	.10	.10	.11
9	.09	.10	.11	.12	.13

To use Table IX–3, add tabular values to the lower sea-level pressure.

EXAMPLE.

Suppose that the station pressure is 24.24 and mean temperature is 40°. Then, entering the table at 40°, the sea-level pressure is between the values corresponding to station pressures of 24.20 and 24.30. These values, from Table IX–2, are 31.02 and 31.14. Their difference is 0.12. The hundredths figure in the actual station pressure of 24.24 is 4. Going down the first column of Table IX–3 to 4, and across to the 0.12 column, we find the value 0.05. This is added to the lower sea-level pressure: 31.02 + 0.05 = 31.07, which gives the final answer.

FURTHER EXAMPLES.

1. Station pressure = 23.83. Mean temperature = 48.6°. (Nearest table temperature is 48.) Then sea-level pressure is between 48° values for 23.80 and 23.90, or between 30.43 and 30.55. Their difference is 0.12.

Using Table IX–3, we find that we must add 0.04 to the lower sea-level pressure (30.43), so the final sea-level pressure is *30.47*.

2. Station pressure = 24.48. Mean temperature = 40.8°.

By proceeding as in example 1, the sea-level pressure is found to be 31.37.

Fixed Cistern Barometers

Some barometers in use are of the "fixed cistern" type, Fig. 68. In these barometers there is no ivory point to which the mercury level must be adjusted. Instead, in the design of the instrument, the relation between the inside areas of the cistern and of the barometer tube is carefully worked out, and in taking a reading a "correction for capacity" is applied. Since this correction for capacity remains the same in a given barometer as long as the quantity of mercury in the barometer remains unchanged and the inside dimensions of the barometer and cistern remain the same, the correction for capacity can be incorporated in the calibration of the barometer scale (all divisions are shortened by a calibrated amount).

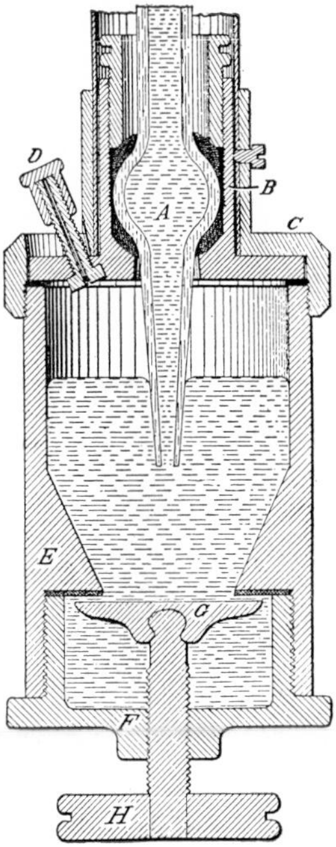

Aneroid Barometers

The *aneroid barometer* is so named because it is, literally, "without liquid." Basically, it consists of two parts—a thin metal box, evacuated partially or wholly of air, and a strong spring. If the box is thin and flexible, for practical purposes we may say that the spring measures the atmospheric pressure.

The motion of contraction or expansion of the spring due to changes in atmospheric pressure is transmitted to a pointer. The atmospheric pressure can then be read from the position of the pointer or needle on a calibrated scale (Fig. 69).

U.S.W.B. photograph.

Fig. 68. Fixed cistern barometer: *G* valve seal; *H* milled head; *E* cistern; *F* bore of cistern; *A* mercurial tube; *B* packing gland; *C* collar; *D* vent. This type is not in general use, but it illustrates the fixed cistern principle.

Reading the Aneroid Barometer [10]

Most aneroid barometer dials are graduated in 0.02 of an inch. It is therefore very simple to read them to the nearest hundredth of an inch. Before making a reading, tap the face of the barometer gently with finger or pencil eraser to make sure that the pointer is not sticking.

The aneroid barometer reading should be compared frequently with a mercurial barometer reading made at the same time and

FIG. 69.   Aneroid barometer.

place, and a correction based on the difference in readings should then be applied to subsequent aneroid reading until a new comparison is made.

ANEROID BAROGRAPH

The aneroid barograph adapts the principle of the aneroid barometer to the purpose of making a continuous record of atmospheric pressure. In place of a single thin metal box as described for the aneroid barometer, a number of metal boxes from which air has been removed are connected. A pen arm replaces the pointer. The pen at the end of the pen arm traces on graph paper attached to a clock-driven rotating cylinder a line representing the atmospheric pressure.

FIG. 70.   Aneroid barograph.

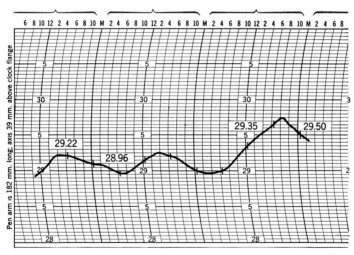

FIG. 71.   Barogram of a one-to-one scale barograph.

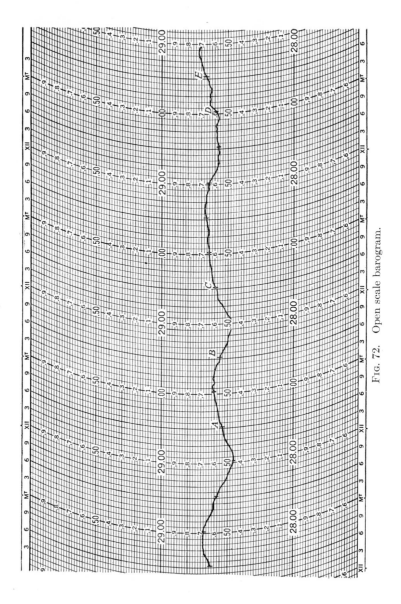

Fig. 72. Open scale barogram.

1. On charts for the one-to-one scale barograph each of the smallest vertical spaces represent 0.05 inch. Sample readings are shown in Fig. 71.

2. On charts for the open-scale barograph each of the smallest vertical spaces usually represents 0.02 inch.

Thus, in Fig. 72, point *A* represents 28.54 inches, point *B* represents 28.56 inches, point *C* represents 28.59 inches, point *D* represents 28.59 inches, and point *E* represents 28.66 inches.

Chapter X

PRECIPITATION [17]

GENERAL

Moisture that falls to earth is called *precipitation*. It may be in any form, such as rain, snow, hail, dew, and frost. The depth of water or melted snow that falls on a level surface in a given period of time, assuming that it were allowed to accumulate, is the basis for measurement. The most common measuring device is the rain gage.

The rain gage is the oldest of meteorological instruments because, in principle, it is the simplest. The earliest record of a series of rain gages, dated about A.D. 1442, ascribes their use to the Koreans. They were in "the form of heavy bronze cylindrical vessels about 15 inches high and 7 inches in diameter, and were set up on stone blocks which were recessed to receive them." Curiously enough, the first rain gage of scientific value designed in Europe was a recording gage that emptied itself when filled to a certain height. It was made by Sir Christopher Wren in 1662. The designer of the first gage in use in America is not known, although it is known that the graduations of the gage were to the thousandth part of an inch.

The accurate measurement of snowfall has been one of the most difficult problems in meteorology, for the snowfall gage must not be affected by high winds that often accompany the snow and tend to destroy the value of the record. A satisfactory yet economical recording or automatic gage is a problem which has not yet been solved. The best exposure of gages is a controversial question that is still the subject of many experiments.

Most regular United States Weather Bureau stations have in operation weighing or tipping-bucket gages that automatically record the depth of rainfall on a clock-driven graph.

STANDARD RAIN GAGE

The 8-Inch Rain Gage

The 8-inch rain gage is so called because the inside diameter of the receiver *A* in Fig. 73 is exactly 8 inches.

The receiver is provided with a funnel that conducts rain from the receiver into the cylindrical brass measuring tube *C*. The cross-sectional area of the receiver is just 10 times the cross-sectional area of the measuring tube. Therefore, the actual depth

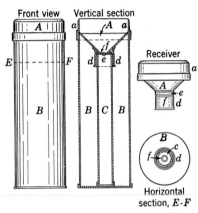

FIG. 73. Standard 8-inch rain gage: *A* receiver; *B* overflow cylinder; *C* measuring tube; *a* cylindrical portion of receiver; *d* sleeve or receiver; *e* inflow hole; *f* funnel.

of rainfall is increased tenfold on being collected in the narrow measuring tube. If water is 10 inches deep in the measuring tube, the actual rainfall will have been only 1 inch, or if the water in the tube is only 0.1 inch, the rainfall will have been 0.01 inch.

For measuring the depth in the measuring tube, a thin cedar stick, graduated in inches and tenths of an inch, is used. As stated above. 0.1 inch graduation represents 0.01 inch of rainfall.

The measuring tube is just 20 inches in height inside. Therefore, exactly 2 inches of rainfall will fill it. Rainfall in excess of this amount will overflow into the overflow cylinder, which is 8 inches in diameter. Whenever the rainfall exceeds 2 inches between measurements, the excess must be poured into the inner tube for measurement. The brass tube must be securely anchored, and the rainfall must be poured very slowly to prevent splashing or upsetting the tube. If there is considerable rain in the overflow cylinder, it is theoretically correct to pour the tube full each time. In actual practice this may lead to spilling, and it is best not to fill the tube completely but record exact measurements of the amount in each partially filled tube and then total.

The standard 8-inch gage has the greatest capacity of all the United States Weather Bureau gages, holding slightly less than 23 inches.

Measuring Rainfall.

Insert the measuring stick slowly into the gage through the small hole in the funnel. When the stick touches the bottom of the measuring tube, hold it there for one or two seconds, then withdraw it and examine it to see at which division the top of the wet portion comes. The numbering of this division, as stamped on the stick, gives the actual depth of rainfall. If the top of the wet portion comes between two divisions, use the Weather Bureau rule for disposal of decimals. If the depth is less than 0.005 inch, it is called a "trace." If it is 0.005 inch or more, up to but not including 0.015, 0.01 inch is recorded. The rain gage should be emptied after each observation.

MEASUREMENT OF SNOWFALL

Measurement of Depth of Snow on Ground

Three or more points where the depth of snow appears representative of the fall over the surrounding area may be selected, and the depth measured with an ordinary ruler or yardstick. It is generally a good idea to take some straight-sided object, such as a shovel, and carefully slice out a portion of the snow so that a good cross-section view of it may be obtained. The recorded depth of snow should be the average of several measurements. For drifted snowfall, as many as ten measurements would be desirable, but not less than three measurements should be made to obtain a representative value.

The place selected for measuring snow depth should be representative of the area, not only with respect to drifting, etc., but also with respect to exposure. For instance, a place that is *always* in the shade should not be chosen, since the snow will disappear more slowly in such a place. All snow-depth measurements should be made near the same place as that originally selected, so that the measurements over several days can be correlated.

Measurement of Melted Snow (Precipitation)

In the winter, when snowfall may be expected, the receiver and measuring tube are removed from the rain gage, leaving only the 8-inch overflow cylinder exposed to serve as a gage for both liquid and frozen precipitation. The overflow cylinder in which snow, frozen rain, rain, or sleet has collected is best taken inside for measurement.

1. Measure an amount of lukewarm tap water in the measuring tube and add it to the snow or ice in the overflow. Slosh it around until all the snow or ice is melted. In the case of frozen rain, melt the ice clinging to the inward sides of the overflow cylinder also.

2. The mixture of tap water and snow water is then measured in the measuring tube by means of the cedar stick. The amount of water that was added for melting the snow or ice must be subtracted from the total amount to give the actual amount of precipitation.

Alternative Method for Measurement of Melted Snow

When the amount of snow collected in the overflow does not seem representative of the total snowfall since the last observation, select a place where the snowfall is as little disturbed as possible and cut out a section with the 8-inch overflow can of the rain gage by slowly sinking the inverted overflow can through the snow and sliding a sheet or thin board under the mouth of the can. (It is a good idea to leave such a board exposed on the ground. Then, when it is covered with snow, simply sink the overflow can through the snow down to the board and then lift up the board together with the overflow can.) The 8-inch cylindrical section of snow may then be melted and measured. At observatories where delicate spring balances are available (sensitive to within 1/4 ounce) the snow need not be melted, but may be weighed, and the depth of water corresponding to its weight may be found by means of Table X–1.

TABLE X–1 (for 8-inch gage, only)
(1 pound equals 0.55 inch)

Weight, pounds	0.00	0.01	0.02	0.03	0.04	0.05	0.06	0.07	0 08	0.09
	Inch	Inch	Inch	Inch	Inch	Inch	Inch	Inch	Inch	Inch
0.0	0.00	0.01	0.01	0.02	0.02	0.03	0.03	0.04	0.04	0.05
0.1	.06	.06	.07	.07	.08	.08	.09	.09	.10	.10
0.2	.11	.12	.12	.13	.13	.14	.14	.15	.15	.16
0.3	.17	.17	.18	.18	.19	.19	.20	.20	.21	.22
0.4	.22	.23	.23	.24	.24	.25	.25	.26	.26	.27
0.5	.28	.28	.29	.29	.30	.30	.31	.31	.32	.33
0.6	.33	.34	.34	.35	.35	.36	.36	.37	.38	.38
0.7	.39	.39	.40	.40	.41	.41	.42	.43	.43	.44
0.8	.44	.45	.45	.46	.46	.47	.47	.48	.49	.49
0.9	.50	.50	.51	.51	.52	.52	.53	.54	.54	.55

Wind Shields for Precipitation Gages

Many gages, particularly of the weighing type, are fitted with a wind shield. The shield is usually of the Alter type, which is a funnel-shaped construction of slats, surrounding the opening of the gage.

The shields are of great importance in catching representative amounts of precipitation, especially snow. When a wind is blowing, and a shield is not used, the sides of the gage cause wind eddies to form around and over the top of the gage, so that precipitation is blown away from the opening. If dry, light snow is falling, these eddies may even blow snow out of the gage. Unshielded gages may catch as much as 10 per cent less than shielded gages.

Automatic Recording Rain Gages

The *tipping-bucket gage* furnishes a means of automatically registering each hundredth of an inch of rain as it falls.

The receiver has a diameter of 12 inches and is attached to a funnel. The funnel directs the stream of collected rain water over the center of the tipping-bucket bearing. As the section of the tipping bucket under the stream fills, it tips, causing a contact to close an electric circuit. This actuates a pen on a register. The tipping bucket is so designed that 0.01 inch of rain will cause the

Photograph by Julien P. Friez and Sons, Inc.

FIG. 74. Alter wind shield for reducing wind effects at the rain gage.

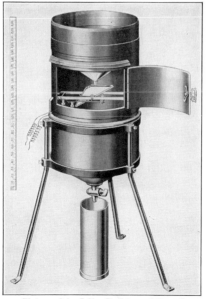

Photograph by Julien P. Friez and Sons, Inc.

FIG. 75. Tipping-bucket rain gage.

bucket to tip. As one section tips, the opposite section is brought under the funnel opening. Thus, every time 0.01 inch of rain accumulates, a contact is made and recorded on the register.

The rain emptied by each tip of the bucket is collected at the bottom of the gage. At each observation, this water is drawn off

U.S.W.B. photograph.

Fig. 76. Weighing-type rain gage.

through a stopcock at the bottom into a special 10-inch measuring tube provided with the gage. After each observation the bucket is tipped ten times to record the time of rainfall measurement. The special 10-inch-deep measuring tube must be used only with the tipping-bucket gage. If for any reason the narrower measuring tube belonging to the 8-inch gage is used, the apparent depth in it must be *divided* by 2.25. The measured amount is used to check the amount recorded on the register. During heavy rain the bucket cannot tip often enough to record the true amount of rainfall. On the other hand, if less than 0.01 inch of rain falls, the tipping bucket will not operate. If a tipping bucket has not tipped, but is more than half full, the amount in the bucket may be regarded as 0.01 inch of precipitation.

The tipping-bucket gage can be damaged by allowing precipitation to freeze in the buckets. When freezing is likely to occur, the tipping buckets and cylindrical portions of the gage should be removed or carefully covered. However, they should be replaced whenever a rainfall at above freezing temperatures is impending.

The *weighing-type rain and snow gage* weighs the rain or snow as it falls and automatically registers the depth of precipitation corresponding to the weight. A pen attached to a spring scale makes a continuous record on a clock-driven chart. At some United States Weather Bureau installations the gage is placed on a tower and fitted with an Alter-type wind shield, to improve the exposure.

USE OF CALCIUM CHLORIDE

Calcium chloride may be used in rain gages at below freezing temperatures to keep the precipitation in liquid form. The concentration of the calcium chloride solution must be kept high enough to prevent freezing at whatever temperatures may be expected. The basic concentration is one volume of calcium chloride to one volume of water. The solution is poured into the bucket, and a film of oil poured over the solution to prevent evaporation.

As precipitation accumulates in the bucket, the solution is, of course, weakened. In order to keep the solution strong enough to prevent freezing, the solution must be recharged or additional calcium chloride added. The instructions on this issued by the United States Weather Bureau Hydrologic Supervisor at Cincinnati, Ohio, for his region are quoted: [21]

TABLE X–2

Temperature Range Expected	Add a Carton of CaCl₂ When Chart Reads:	Add a Second Carton of CaCl₂ When Chart Reads:	Empty and Recharge When Chart Reads:
+32° to +10°	5½ inches	....	10 inches
+10° to −10°	4½ inches	....	8 inches
−10° to −30°	2½ inches	5½ inches	8 inches
−30° to −60°	2½ inches	5 inches	7 inches

Chapter XI

WINDS-ALOFT OBSERVATIONS [13]

GENERAL

Wind direction and speed above the surface of the earth are measured by observing the path taken by a freely ascending balloon. The visual method employs either single or double theodolites to observe horizontal (azimuth) and vertical (elevation) angles at 1-minute intervals. The ascensional rate of the balloon is assumed constant when only one theodolite is used, and the path of the balloon may be plotted on a large protractor. At night a small lantern or flashlight is attached to the balloon.

The radio method employs a small radio transmitter attached to the balloon and a ground radio direction-finding set to follow it. The radar method uses a target below the balloon which the ground radar follows to give slant range and azimuth and elevation angles as the balloon ascends.

All these methods require a balloon which is moved by the wind as it ascends through the atmosphere. Observation of the position of the balloon and its height at the beginning and end of a time interval gives the necessary data with which the mean wind direction and speed through the layer may be computed. The balloon and its attachments are so light that they respond immediately to changes of moving air layers through which they pass.

SELECTION OF SITE FOR SINGLE THEODOLITE OBSERVATION

The observation should be made from a point from which an unobstructed view of the entire horizon may be obtained. Orientation points should be located nearby. The exact direction as measured from true north from the theodolite position to the orientation point should be determined when the observatory is established.

Balloons

Pilot balloons, as the balloons used in these observations are called, are made of rubber or Neoprene and usually weigh 30 grams or 100 grams. The United States Weather Bureau balloons are generally filled with helium, although in outlying stations hydrogen is generated for this purpose. The free lift given to the balloon to provide the necessary buoyancy is as follows:

Free Lift

	Helium	*Hydrogen*
30-gram balloons	154 grams	140 grams
100-gram balloons	503 grams	450 grams

Hydrogen has more buoyancy; therefore, less "free lift" is required.

Theodolites

The theodolite is a specially designed instrument similar in many respects to a surveyor's transit. A small telescope is

U.S.W.B. photograph.

Fig. 77. Theodolite.

mounted in such a manner that it can be turned on both a horizontal and a vertical axis. A vertical circle graduated in degrees is provided for determining the elevation angle above the horizon when the balloon is sighted. A horizontal circle, similarly graduated, is provided for reading the azimuth angle from true north or the direction of the balloon, with reference to the theodolite, as it is observed in free flight. Tenths of degrees may be read from vernier scales.

The rising balloon is observed continuously during flight, and, at 1-minute intervals, elevation and azimuth angles are read and recorded.

COMPUTATIONS

Tools necessary to make computations of wind direction and speed at standard levels comprise a plotting board or protractor and a slide rule or tables. The United States Weather Bureau

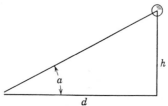

FIG. 78. Vertical triangle for computation of "distance out." $d = h/\tan a$.

uses a graphical method of computation, which requires that the track or path of the balloon be plotted on a protractor, using the azimuth angle and a computation of the horizontal distance of the balloon from the station at the end of each 1-minute interval. For standard balloons, tables are available of horizontal distance at each minute period for various values of the elevation angle.

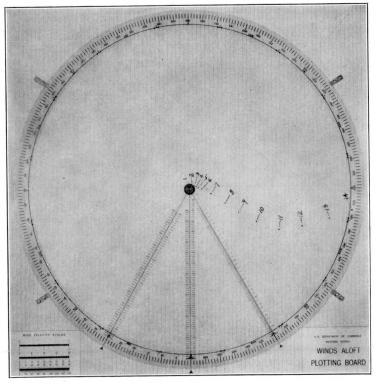

FIG. 79. Protractor for computation of wind speed and direction in the upper air.

GRAPHS

Direction and speed of winds aloft are graphed as shown in Fig. 80. A detailed description of double theodolite computations and the tail method of winds-aloft computation will be found in United States Weather Bureau Circular O. These methods are in such restricted use as not to justify a description of them here. Description of winds-aloft observations that utilize electronic means will be found in Chapter XII.

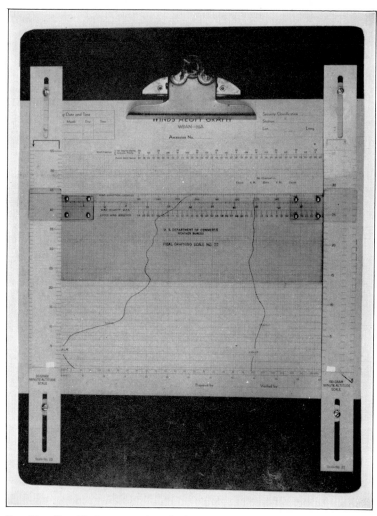

FIG. 80.   Graphical presentation of wind direction and speed in the upper air.

Chapter XII

ELECTRONIC METEOROLOGICAL OBSERVATIONS

During the last half century meteorologists have been striving to attain better methods of determining the state of the upper atmosphere, and to this end meteorographs, which are instruments for measuring and recording on a chart pressure, temperature, and in some cases relative humidity, were developed. Since these instruments were carried aloft by a free balloon, availability of the recorded data depended on recovering the instrument after its descent. In the United States approximately 80 per cent of the instruments released were later recovered, but the lapse of time between the release of the meteorograph and its subsequent recovery was always such that the data could not be used for current forecasts. To eliminate this lapse of time and to insure recovery of the instruments the Weather Bureau developed kites to carry them aloft. The kites were later superseded by airplanes. Both kites and airplanes, however, had the disadvantage of being unusable during stormy weather, when information on the condition of the upper atmosphere was needed most urgently.

The problem presented by this situation was solved with the development of an inexpensive radiometeorograph, which is carried aloft by a free balloon while transmitting temperature, pressure, and humidity data to the ground by radio signals. During the war, radio direction-finding equipment, operated from the ground, was developed for tracking the position of the radiometeorograph during its ascent. It thereby became possible to determine the direction and speed of the winds aloft at all levels between the point of release and the bursting point of the balloon. Later developments permitted the determination of certain large-scale weather characteristics by a modification of the radio direction-finding principle (sferics), or alternately by the principle of radio direction and ranging (radar).

RADIOSONDE OBSERVATIONS

For convenience, radiosonde observations are customarily referred to as raobs. The observations are taken with a radiosonde, which consists of a thermometer, a hygrometer, and a radio transmitter, all contained in a small, light-weight box. As the radiosonde is carried up through the atmosphere by a helium- or hydrogen-filled balloon, radio signals are transmitted to the ground receiving station, where they are automatically recorded. When the balloon bursts, the radiosonde descends to the surface of the earth on a small parachute.

THE RADIOSONDE

Radio Transmitter

The radiosonde transmitter, one type of which is shown in Fig. 81, emits a radio signal at a frequency of 72.2 megacycles.

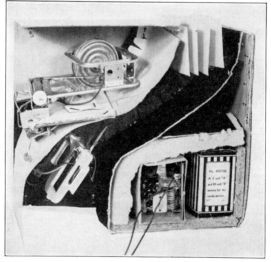

FIG. 81. Radiosonde. (Cutaway shows curved air duct with meteorological elements on the left side and radio transmitter and battery on the lower right.)

The signal is modulated by varying the resistance in the meteorological control circuits by means of resistors sensitive to changes in temperature and relative humidity and by two additional fixed resistors. The fixed resistors are in circuits termed the high and low reference circuits, respectively. Figure 82 shows schematically the meteorological control circuits of the two models of modulated audiofrequency radiosondes now in general use in the United States.

Baroswitch

The baroswitch, shown in Fig. 82, has two functions in the radiosonde: (1) to indicate pressure values during the sounding,

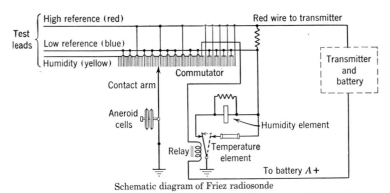

Schematic diagram of Friez radiosonde

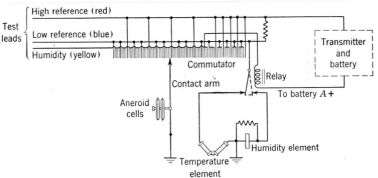

Schematic diagram of W.I.T. radiosonde

Fig. 82. Schematic diagram showing commutator operated by aneroid cells as a baroswitch.

and (2) to switch into the control circuit in a definite order the temperature, humidity, low reference, and high reference resistors. One side of the pressure diaphragm is fixed to a rigid support; the

other side of the distending diaphragm engages a contact arm
through a suitable linkage. As the radiosonde ascends through

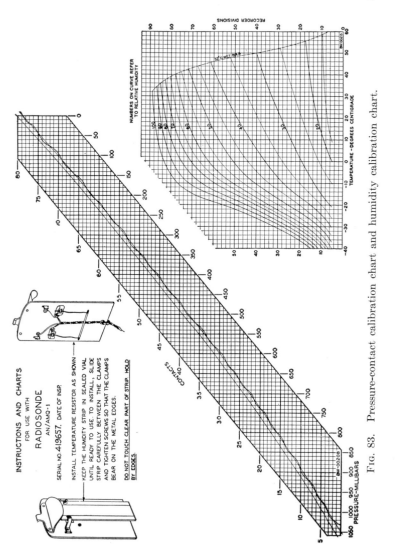

FIG. 83.   Pressure-contact calibration chart and humidity calibration chart.

levels of decreasing atmospheric pressure, the distending diaphragm of the baroswitch causes the contact arm to move across a
commutator.

The baroswitch commutator consists of either 80 (Friez) or 95 (Washington Institute of Technology) metallic segments separated by dielectric material. One metallic segment and the succeeding adjacent nonconducting segment comprise one "contact." The 80 (or 95) contacts are correlated with the indicated pressure values in such a manner that, when the number of the contact is known, the corresponding indicated pressure can be determined from a calibration chart as shown in Fig. 83.

By referring to Fig. 82, it can be seen that, when the point of the contact arm rests on any one of several metallic segments, a relay is energized so that the humidity resistor is connected into the meteorological control circuits. The other metallic segments are connected to the high and low reference circuits. When the contact point rests on a nonconducting segment, the temperature resistor is in the control circuit.

Temperature Element

The temperature resistor, or element, is made of a ceramic material, the resistance of which increases as the temperature decreases. Figure 84 shows the type of temperature element now in general use.

Fig. 84.   Temperature element (ceramic).

For each model of radiosonde, a temperature evaluator is provided. This evaluator is a two-scale slide rule designed to convert the recorded temperature ordinate into degrees Centigrade. As indicated in Fig. 85, one scale of the evaluator represents the temperature ordinates, the other Centigrade temperatures. During a prerelease check of the radiosonde, known as the "baseline check," the temperature evaluator is set with the instrument shelter temperature opposite the corresponding recorded temperature ordinate. From this setting, the temperature for any level of the sounding can be determined by reading the temperature value opposite the temperature ordinate recorded for the level.

The term ordinate is used as equivalent to the terms "recorder ordinate," "chart division," and "frequency division" which are found on the several types of evaluators and calibration charts used.

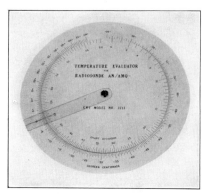

FIG. 85.   Temperature evaluator.

Humidity Element

The humidity resistor or hygrometer element consists of a chemically coated plastic strip with metallized edges. The resistance across the chemical film changes with variations in the relative humidity and the temperature of the air in which it is exposed. The effect of temperature is eliminated by means of a special graph, from which the relative humidity of a given level of the sounding can be determined as a function of both the temperature and the humidity ordinate. Figure 83 shows a radiosonde calibration chart containing both the humidity evaluation graph and the pressure calibration curve.

Ventilation Chamber

The temperature and humidity elements are installed in the ventilation chamber of the radiosonde. The elements are surrounded by a cylindrical shield designed to reduce the effects of radiation and insolation on the measurements. The methods of mounting the temperature and humidity elements are shown in Fig. 81.

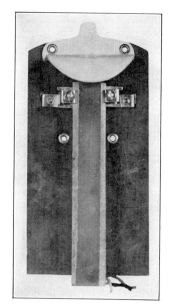

FIG. 86.   Humidity element. Glass strip coated with lithium chloride solution.

The Ground Equipment

General

The radiosonde ground equipment consists of an antenna, a short-wave radio receiver, an electronic frequency unit, and a recorder. A voltage stabilizer is added if improved regulation is necessary because of fluctuations in the available power supply.

Fig. 87. Radiosonde receiver and recorder (radio receiver and frequency meter).

These units are used in the process of receiving and recording the radiosonde signals.

Antennas

The dipole and wire-doublet antennas are the two types of radiosonde antennas in general use. The standard vertical dipole antenna consists of a metal rod and a metal skirt, each approximately one-quarter wavelength long. A gas-filled or a solid coaxial cable transmission line is used with this type of antenna to reduce loss of signal strength. The wire-doublet antenna, consisting of two legs, each of which has a length of approximately one-quarter wavelength, is used for standby or emergency purposes. Coaxial cable transmission line also produces better results with this antenna, although twisted-pair telephone wire may be satisfactory for the transmission line provided its length is kept at the minimum to avoid loss of signal strength.

Receivers

Superregenerative and superheterodyne receivers with especially designed audio-amplifiers are used for radiosonde reception on the 72.2-megacycle frequency. Superregenerative receivers are regarded as the better type for radiosonde observations, since they require less frequent tuning than the superheterodyne type and permit the observer to spend more time in evaluating the data as the record is made. However, at stations having considerable interference, the superheterodyne receiver usually gives better results, since it provides greater selectivity.

Electronic Frequency Units

All the common electronic frequency units embody the same operating principle. A pulsating direct current proportional to the frequency of the applied alternating-current voltage (input signal) operates a visual meter and a recorder. A current output of approximately 500 microamperes is required for full-scale deflection of 100 ordinates. The several types of frequency units in service differ principally in mechanical design. Elaborate voltage regulation is required so that the output current is controlled by the input signal without being affected by variations in the power supply.

Recorders

The two radiosonde recorders in general use are the microammeter and potentiometer types.

MICROAMMETER RECORDER. The microammeter recorder is designed to register the position of a microammeter pointer. The position of the pointer depends on the current output of the electronic frequency unit. A photoelectric scanning device causes an impression to be printed on the recorder chart paper at a point corresponding in value to the position of the pointer.

POTENTIOMETER RECORDER. In the potentiometer recorder, the output of the electronic frequency unit is balanced against an automatically operated slidewire potentiometer. The balancing mechanism moves a pen to the point on the recorder chart paper corresponding to the adjustment of the potentiometer.

Because of the detail and the frequent changes in instruments, *United States Weather Bureau Circular* P should be referred to for further information on this type of observation.

RAWINSONDE

One of the most notable contributions to aerial navigation and upper-air meteorology during the war was the development of radio direction-finding units that could be used in conjunction with balloon-borne radiosonde transmitters to obtain winds-aloft observations at high levels during all weather conditions. These observations are termed "rawinsondes," a phrase contraction of "wind and radiosonde." The direction-finding equipment was developed by the Army Signal Corps and the Farnsworth Television and Radio Corporation, and it is now being made available to the Weather Bureau for its upper-air program. The principal components of the new equipment are the radio receiver indicator with azimuth and elevation radar-type oscilloscope and antenna, designated as the SCR–658 type. The Weather Bureau has been operating 76 radiosonde stations in the United States and its possessions, and on ships in the Atlantic Ocean, where upper-air observations of temperature, pressure, and humidity are made by means of a balloon-borne radio-meteorograph called a radiosonde. This instrument transmits pressure, temperature, and humidity signals on a frequency of 72.2 megacycles to a radio receiver on the ground. The signal is automatically recorded, and values of the upper-air elements are obtained to the bursting point of the balloon.

The new program involves a radiosonde operating on a frequency of 403 megacycles, which is suitable for radio direction-finding. The antenna of the SCR–658 may be rotated through the vertical and horizontal axes so that maximum signal strength will be indicated by the oscilloscope when the antenna array is pointed directly at the ascending radiosonde. Elevation and azimuth angles may then be read on indicators at 1-minute intervals. Direct computation of the altitude of the radiosonde may be made by pressure-temperature relationships, utilizing the values transmitted by the radiosonde during its flight.

In actual practice, two observers working as a team are required to make the observation, compute the values, and code the data for transmission by teletype and radio at the scheduled times. One observer acts as the SCR–658 operator, and he is in general charge of the assembly of the radiosonde train, and of the actual operation of the direction-finding equipment. A radiosonde observer operates the radiosonde ground equipment, and he computes the pressure, temperature, altitude, and humidity data as they are received on the recorder. The first observer also computes the winds aloft. Azimuth and elevation angles are recorded each minute by the operator of the SCR–658 set; altitudes are computed and given to him by the radiosonde observer as the flight progresses. With these data he computes the position of the balloon with respect to the station at each minute interval, and then obtains the wind direction and speed through the various levels of the flight.

RADAR WIND FINDING

Radar is a phrase contraction for "radio direction and ranging." It represents a development made during the second World War, originally for gun laying and bombing purposes, that led directly to a meteorological application.

Radar propagates very short-wave bursts and receives back echoes from objects in the path of the bursts. The time difference between emission of the burst and reception of the echo is proportional to the distance of the object producing the echo. The direction of the object can be determined from the orientation of the antenna array.

In practice, a free balloon is released with an attached target that is tracked by the radar ground equipment in a manner similar to that of a theodolite. The range, and the elevation and azimuth angles, are used to compute by graphical means the wind vectors pertaining to all levels between the point of release and the bursting point of the balloon. The accuracy of the radar wind finding method is as good as or better than the single theodolite method.

Storm Detection

Radar is also applied in meteorology to the discovery and tracking of areas of clouds and precipitation. By using a 3-cm or a 10-cm wavelength and a plan position indicator scope, a "map view" of storm areas may be continuously observed at radar stations. Usually an area of 150 to 200 miles in radius is covered. This gives the meteorologist the first moving weather map representing actual conditions over a large area.

OTHER DEVELOPMENTS

Sferics

The term "sferics" is a contraction of the word "atmospherics." It is applied to a system of weather observation that makes use of the electrical characteristics accompanying many atmospheric phenomena. Sferic determination of electrically active areas requires the use of two stations separated by distances of 100 miles or more. Each station is equipped with a radio receiver operating from a loop antenna. Each antenna has two components, one for response to signals coming from a north-south direction and the other for signals coming from an east-west direction. When an electrical discharge is picked up, a cathode-ray tube indicates its direction from each of the sferic stations. Communcation between the stations is immediately established, and the bearings of each on the discharge are exchanged. Since the distance between each two stations is known, it becomes a matter of simple triangulation to compute the distance and location of the discharge.

Automatic Weather Station

Meteorological data are frequently required from areas that are not suitable for habitation by weather observers. To satisfy the requirement automatic weather stations have been developed that have worked satisfactorily on mountain tops, remote and sometimes wave-washed islands, swamps, and even on buoys anchored in the ocean. These stations have been especially useful under hurricane conditions, for their power supply is self-contained and the stations continue to function long after weather observers would have taken shelter.

The automatic weather stations used by the United States Weather Bureau transmit, on a frequency of approximately 3 megacycles, values of atmospheric pressure, wind direction, and wind speed. The data can be transmitted approximately every 30 seconds. Since they are powered by a storage battery, however, the transmitters are usually activated by means of a program clock only eight times a day. Under normal conditions of operation the automatic weather station will not require attention more often than once every five months, when the storage batteries must be recharged. With slight modification the instrument could be used to transmit rainfall, temperature, and humidity data.

The transmitted signals of the automatic weather station are usually recorded automatically, the receiver and recorder being turned on simultaneously with the transmitter. The receiving station itself may be automatically operated and so arranged that the received signals are retransmitted by land line to a weather office. The use of a receiving station remote from the weather office is occasionally necessary to insure receipt of dependable signals from the transmitter.

Chapter XIII

AN IMPROVISED WEATHER STATION *

GENERAL

In setting up an improvised weather station, observe all precautions used in setting up a general meteorological station. The outside equipment should not be too close to obstacles; in general, it should be as far away from obstacles (trees, houses, etc.) as the obstacles are high. Extremes of topography such as dips and hollows, hilltops, and mountain crests should be avoided. For all instruments for the measurement of temperature and humidity a proper shelter has to be provided. The objectives of the shelter should be: to protect the instruments from direct and reflected solar radiation, to prevent precipitation from striking the instruments, and to allow free circulation of air around the instruments. Before making a thermometer shelter it may be helpful to study Figs. 88 and 89. Note that in many cases a steep roof supported by four legs and a platform for the instruments below will suffice to fulfill the requirements of a meteorological shelter.

Ordinarily, improvised instruments will be suitable for direct reading only, yet the barometer, hygrometer, and thermometer described below can be converted into recording instruments if a clock is available from which the glass and the hands have been removed, and a flat round disk (top of a tin can) has been fastened to the axis of the hour hand. A simple drum from a tin can may also be placed on the axis, provided the clockwork is strong enough to move it. On the disk a piece of smooth paper can be fastened by one or two small pins; on a drum the paper can be held in place by two rubber bands. A recording pen fastened to the end of the pointer of the particular instrument will permit a recording of the element in question. The pens at the end of the pointer can be

* From notes by AAFTTC.

shaped from a piece of thin sheet metal. Recording ink can be made readily from a mixture of 2 parts of water and 1 part of

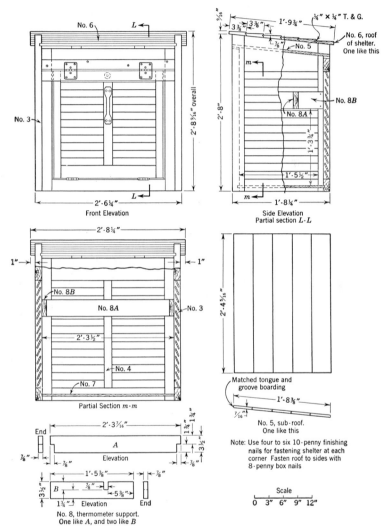

FIG. 88.   Instrument-shelter details.

glycerin, colored by shavings from an indelible pencil. On the whole, home-made recorders are only as good as their clockwork. They also depend on the quality of paper. Further, the pen arms

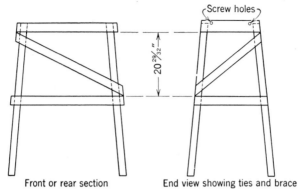

Front or rear section — End view showing ties and brace

FIG. 89. Support for instrument shelter.

FIG. 90. Simple thermometer shelter.

must be constructed from resilient metal to keep the pen on the paper without developing too much friction.

The instruments to be described in the following paragraphs will in general require only the most common types of materials, which are likely to be available almost anywhere, like tin cans, pieces of wood, bolts, and nails. The only tools necessary for building the instruments are usually tin shears, file, hammer, pliers, and soldering iron. Suggestions for using parts from old automobiles and plumbing fixtures that may be on hand are to be found in the individual instrument descriptions. For measuring and calibration purposes, an inch ruler and a protractor are about the only "precision" pieces assumed to be available.

OBSERVATION OF PRECIPITATION

For the observation of precipitation, rain gages are readily built from ordinary tin cans. Figures 91 and 92 show construction sketches and assembly of such a gage. An ordinary rain gage consists essentially of four parts: first and most difficult to build, the funnel; second, the outer receptacle; third, the collecting tube; and fourth, the measuring stick. It is desirable to obtain a reasonable magnification of the accumulated rain so that it can be readily measured. Ordinarily, a magnification of about ten to one is attempted. In order to achieve that, the outer rim of the receptacle funnel should encompass ten times the area of the inner tube. In other words, it is desired that $\pi R^2 = 10\pi r^2$, where R is the radius of the rim of the outer receptacle and r the radius of the inner tube. This indicates that the ratio of the radius of the outer receptacle R to the radius of the inner tube r should be equal to the $\sqrt{10}$, which is equal to 3.16. Roughly speaking, if an outer-receptacle radius of 3 inches is chosen, the radius of the inner tube should be about 1 inch; or, speaking in terms of diameters, one could also choose a can of 6-inch diameter and a piece of tubing of 2-inch diameter for the inner receptacle to achieve the purpose.

The height of the outer receptacle probably should not exceed about the height of three medium-sized tin cans, because it will otherwise be difficult to solder and to get reasonably straight. Most difficult to construct is the funnel part of the rain gage. The procedure is illustrated in Fig. 92, 1. A piece of sheet metal is illustrated there comprising about ⅝ of a circle, which is later soldered together at the ends. This will give about the proper

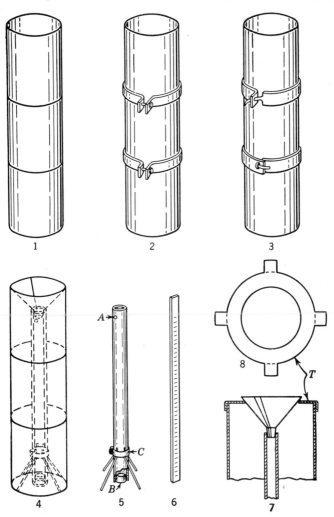

Fig. 91. Rain-gage construction.

slope to the funnel. If a larger portion of the circle is used a funnel with too little slope will result; and if a smaller portion of the circle is used a funnel with too steep a slope will result. The circumference of the ⅝ of the circle should be such as to give the desired outer circumference of the larger receptacle. In other words, it should be equal to $2\pi R$.

The inner tube should be provided with an overflow hole at the top to avert loss of the "catch" if more rainfall occurs than the inner tube will hold. Under these circumstances, the rain will

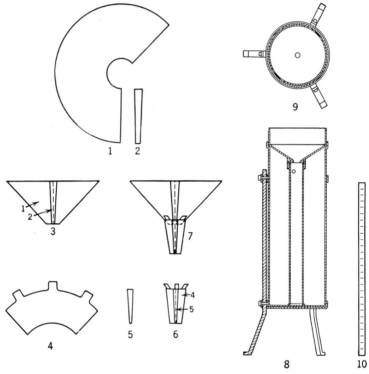

Fig. 92.   Rain-gage-receiver (funnel) construction.

flow into the outer receptacle, and, after the amount in the inner tube has been measured, the portion spilled into the outer receptacle can be poured into the inner tube and then measured.

Any graduated piece of wood can serve for the measuring stick. If the ratio of 1 to 10 is chosen $\frac{1}{10}$ inch on the measuring stick will correspond to $\frac{1}{100}$ inch of rainfall.

In setting up the rain gage one should be careful to have the edge of the funnel in a horizontal position.

If paint is available, it is desirable to have all parts painted inside and outside so that the sheet metal will not rust.

OBSERVATIONS OF THE WIND

For wind observations, it is necessary to measure two quantities: the direction of the wind, and the speed of the wind. Both of them, of course, can be estimated. There are numerous procedures to obtain an estimate of the *wind direction*. The first step is to obtain the approximate astronomical directions at the place where the observation is to be made. This is best achieved by means of star or solar observations. For most meteorological purposes it is entirely satisfactory to have the wind direction correct within about 10° or 20°, so that a rather rough determination of the astronomical direction is satisfactory. Sighting toward the pole star of the northern hemisphere at night should be completely satisfactory to obtain the north-south direction. In daytime, the observation of the shadow of a stick set vertically into the ground should help. To obtain the south direction, for example, the procedure is to follow the shadow of the stick and mark the end of the shadow for a period of time in midday. At solar noon, the shadow of the stick will be shortest, and a rough curve drawn through the end points of the shadow will readily yield the approximate south position of the sun; so that a line drawn from this minimum distance of the shadow curve to the center of the stick will give the north-south line.

Astronomically determined directions are always superior to those obtained from compasses, particularly if the exact declination at the particular locality is not known. In some areas the declinations may be so large, either on account of local anomalies or because of the vicinity of the magnetic poles of the earth, that compass directions are considerably off the true directions.

In estimating the wind direction the face may be turned into the wind until the same force is felt on both cheeks. The direction one faces will then be the direction from which the wind is blowing. This is applicable only in strong winds. In weaker winds it is a more common practice to wet one's finger and hold it up in the wind. The side of the finger that becomes coldest is facing toward the direction from which the wind is blowing, owing to the water's evaporating on the skin. Smoke from a fire or a cigarette may also help in determining the wind direction.

The Beaufort scale of equivalents is still by far the best method of obtaining a satisfactory estimate of the wind speed.

Recording wind vanes and anemometers are instruments that are difficult to construct and that require a precision machine shop, but simple substitute instruments can readily be made. Figures 93 and 94 present some suggestions. If possible, it is

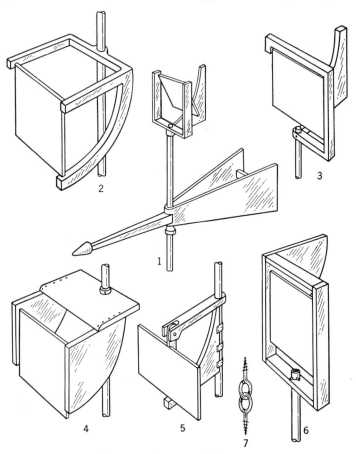

Fig. 93.   Pressure-plate or pendulum anemometers.

advisable to make a combination of wind vane and swinging plate anemometer. Little needs to be said about the construction of a wind vane. It can be built from sheet metal or wood. The steadiest wind vanes are of the split-wing type, an angle of about 22° between the two wings being most advantageous. The weight of the wings should be counterbalanced in front by a counterweight

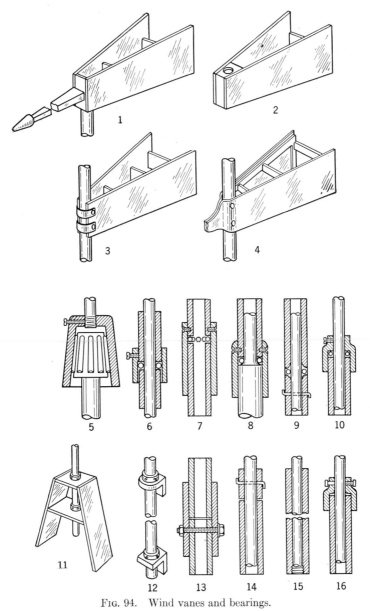

FIG. 94.   Wind vanes and bearings.

as streamlined as conditions will permit. The most difficult portion of the wind-vane construction is the bearing. In order to obtain satisfactory results be sure that the shaft of the wind vane is in a vertical position, and that there is as little friction as possible between the wind vane and the shaft. Numerous suggestions for making bearings of wind vanes will be found in Fig. 94, 5 through 16.

The wind vane is also used to turn the wind plate (pendulum anemometer) into the wind so that the wind will strike the front of the plate squarely. The stronger the wind, the greater will be the angle the plate makes with the vertical. For a sheet-iron plate, 6 by 12 inches and weighing 7 ounces (200 grams), the following table gives the angle the plate will make with the vertical for a given Beaufort force of the wind:

Wind force	0	2	3	4	5	6	7	8
Angle of plate with vertical	0°	4°	15°	31°	46°	58°	72°	81°

In the construction of the plate these dimensions can be readily adhered to even if sheet metal is not available by using wood and giving it the proper weight by fastening a stone or a piece of metal to the back of the plate. The proper weight of the plate, if no scales are available, can be estimated by comparing with the weight of standardized products labeled as to gross weight. Take care that the plate is hung in a vertical position and can swing freely. Numerous suggestions for hinges will be found in Fig. 93.

Observation of Humidity

For measurement of humidity, only two methods are common in meteorology: the psychrometer method, with wet-bulb and dry-bulb thermometers, and the hair hygrometer method. If thermometers are available, there is no difficulty in determining the humidity from wet- and dry-bulb readings, provided that tables for conversion of these readings into relative humidities or dew points are available. Even if only one thermometer should be on hand, it is possible to obtain a humidity measurement by first using it as a dry-bulb thermometer and then putting a wick around it and using it as a wet-bulb thermometer. If no psychrometer tables are on hand, A. J. Turner's formula may be helpful:

$$\text{Relative humidity} = 100 - 3\left(\frac{t - t'}{t}\right)100$$

where t is the dry-bulb temperature and t' the wet-bulb temperature, both in degrees Fahrenheit.   This formula will permit an estimate of the relative humidity within about 3 per cent in the normal ranges of that element.

If thermometers are not available, the observer will have to resort to the construction of a hair hygrometer.   Any type of

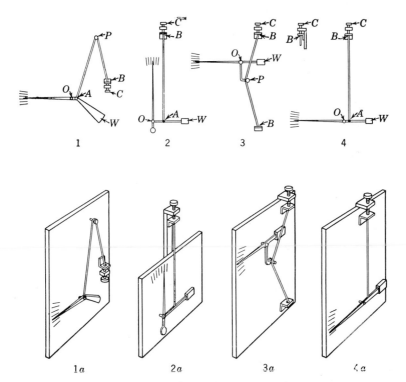

FIG. 95.   Hair-hygrometer construction.

straight human hair will be suitable for the purpose.   Hairs should not be shorter than about 8 inches.   Figure 95 gives some suggestions for the construction of a hair hygrometer.   If the hygrometer is delicately built, a single hair will yield good measurements, provided that it is given sufficient time for adaptation.   This means that the hygrometer should remain in the environment where the measurement is to be made for at least 20 minutes.   For more ruggedly built hygrometers, a bundle of 5 to 15 hairs could be used for the construction.

Before human hair can be used for hygrometric purposes, it has to be cleaned from scales and grease by boiling it in clean water. If pure alcohol, ether, or benzine is available, the hair should be cleaned by soaking it 24 hours in any of these liquids. Thereafter touching the hair with hands or fingers should be avoided as far as possible. Best results with hair hygrometers of the type shown in Fig. 95 have been obtained with about five hairs. Everything, especially the pointer, should be made as light as possible. A thin sliver of wood may often serve for the pointer.

Calibrating the scale of the improvised hair hygrometer is the most difficult procedure that has to precede its use. The upper fixed point, that is, 100 per cent relative humidity, can be found rather easily by placing the whole instrument in a box in which it is surrounded by wet towels. The space will rapidly become saturated with water vapor, and the hair hygrometer should indicate 100 per cent relative humidity. The task is then to find a second point on the scale. Once two points are known the remainder can be determined by the well-known relation between expansion of the individual hair or bundle of hairs and the relative humidity. If sulphuric acid is available, a 50 per cent solution at 68° F. will, for example, yield a relative humidity of 34 per cent in a closed container. A 70 per cent sulphuric acid solution will yield, in an enclosed space, a relative humidity of about 4 per cent. These acid solutions may be available in places where storage batteries are being recharged. If sulphuric acid is used in obtaining calibration points, none should be permitted to fall on the hair, because the acid will impair its usefulness. If ordinary table salt is available, a saturated solution in distilled water will yield, in an enclosed space and after equilibrium has been reached, about 75 per cent relative humidity. The relative change of the length of the hair with relative humidity in terms of the total possible change of the given hair or strand of hairs is shown in the following table:

PERCENTAGE CHANGE OF LENGTH OF HUMAN HAIR IN TERMS OF TOTAL
LENGTH CHANGE WITH VARYING RELATIVE HUMIDITY

% relative humidity	0	10	20	30	40	50	60	70	80	90	100
% of total length change	0	20.9	38.8	52.8	63.7	72.8	79.2	85.2	90.5	95.4	100

If one has obtained any two points on the scale, the proportion of the total length change will be known, and the other points can be obtained by calculation. For example, if one used a 50 per cent sulphuric acid solution yielding 34 per cent humidity to obtain the lower fixed point, it can be seen from the table that 34 per cent would account for about 57 per cent of the total length change possible for human hair. In other words, between the two observed points of 34 and 100 per cent relative humidity would be 43 per cent of the total change that could be observed on the scale of the instrument. This 43 per cent is subdivided in accordance with the ratios shown in the table, and the portion toward the lower relative humidities is extrapolated.

OBSERVATIONS OF PRESSURE

The measurement of the absolute value of atmospheric pressure by means of improvised equipment is difficult to accomplish with accuracy. The reason is that one would need either to build a mercurial barometer or to have means for an accurate calibration of an aneroid barometer. The first requires the availability of accurately drawn capillary tubing and a sufficient amount of mercury, which usually will not be on hand, and both require pressure chambers for calibration which are usually unavailable also. Comparison with a standard barometer over a period of time may be possible. Since measurements of the absolute amount of pressure with improvised equipment seem not to be obtainable, the next best substitute is the measurement of pressure changes. And that, in spite of the fact that the absolute pressures involved in the change may be unknown, is still of great value to the weather forecaster. Two types of instruments to indicate pressure variations will be described here.

One is based on the aneroid barometer; the other is a variometer, which might be called a latter-day descendant of the old familiar New England weather glass. For the aneroid-type barometer, which is shown in Fig. 96, a box evacuated of air is needed, and this may be as simple as a can of vacuum-packed coffee. If such a can is available, the only task of construction is to fasten a pointer to the can to act as an index of any pressure changes that may occur. Various suggestions about the manner in which such a pointer may be fastened to a vacuum packed coffee can are indicated in 6 to 15 of Fig. 96. The disadvantage of this instrument is that the can will react not only to pressure changes but

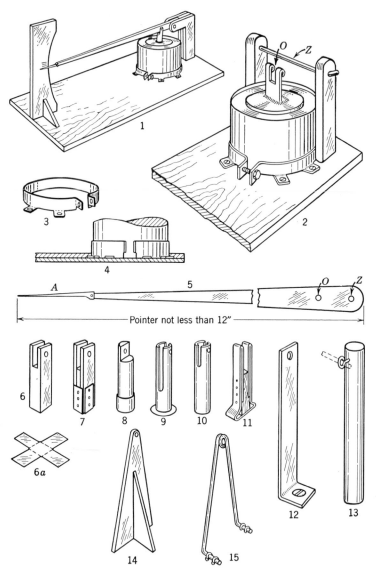

FIG. 96. "Coffee-can" aneroid barometer with components.

also, to some extent, to temperature changes. It has, therefore, to be kept at an even temperature and must particularly be protected against direct solar radiation. In areas where the daily variation of temperature is very large, the can will have to be protected by an insulating box. Yet, generally, the can will react against pressure more than against temperature changes, and thus

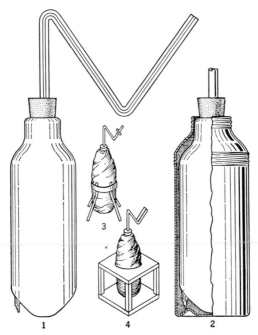

Fig. 97. Vacuum-bottle variometer.

all major and rapid pressure changes will be readily indicated by the instrument.

The second type of instrument is a variometer, which will indicate changes of pressure over a specified interval of time. It may be constructed from a thermos bottle which is sealed off against the outside by means of a tightly fitting stopper of rubber or waxed cork. Through the stopper a piece of glass tubing is snugly fitted, bearing some indicator liquid in a U-shaped extension. One of the arms of the U-shaped indicator tube may bear a scale or an index to facilitate the reading. Any slowly evaporating liquid (such as oil) will serve as indicator. If available, a tube with a relatively narrow bore should be used. Such tubes can be made out of capillaries salvaged from broken thermometers or barometers, but any other piece of glass tubing will do. The wider the bore, the less sensitive the instrument is. A certain amount of control over the sensitivity can be exercised through the slope adopted for the indicator tube.

If the outside pressure falls, the liquid in the indicator will move toward the end of the capillary; if the outside pressure rises, the liquid will move in the capillary toward the side of the flask. Since the enclosed volume of air in the bottle will react not only to pressure but also to temperature, a thermos bottle in which the temperature can be kept relatively even is advantageous. In order to control the temperature inside of the bottle, it is advisable to put a mixture of water and ice into the bottom of the bottle, because the mixture will keep the temperature exactly at the freezing point, and, as long as both water and ice are present, the temperature will be maintained at that point. Whenever the ice is melted, another piece of ice can be added and the temperature brought back to the freezing point. If a vacuum bottle is not available, any ordinary glass bottle will serve the same purpose, although not so well. In any event, it should be insulated against temperature changes as well as possible. This is recommended even with the thermos bottle, which should be wrapped in cotton or cloth and put into a protective box with only the indicator capillary visible. Protection against direct sunlight is essential.

Observations of Temperature

The only type of thermometer that is common in meteorological observatories is a "liquid-in-glass" thermometer. It is impossible to improvise constructions of that type, and therefore the instruments that can be built as substitute equipment for thermometers will have to be metallic expansion thermometers or bimetal thermometers. Drawings showing the construction of these are given in Fig. 98. Drawing 1 of that figure shows a metallic expansion thermometer for which a piece of metallic tubing, of the type frequently found in engines, or any stiff long rod, can be used. This piece of tubing or rod is fastened to one side of and against a heavy block of wood or stone, so that it can expand only at the other end if the temperature changes. At that end the pointer is pressed against the tubing by means of a little steel spring, and thus a thermometric reading can be obtained. Calibration of such a piece of equipment without thermometers available for comparison is rather difficult, although wherever water and ice are available at least the freezing point (32° F.) can be established on the scale. A second approximate fixed point can be obtained by warming the temperature-sensitive part of the element with

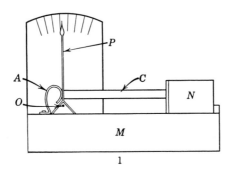

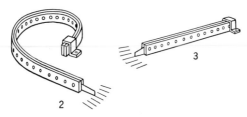

FIG. 98.   Bimetallic thermometer.

the hands, the inside palm temperature being usually about 90° F. Subdivisions are linear on this instrument. The sensitivity of the instrument will depend, of course, in part upon the length of tubing used for the measuring body, which should not be too small. Eight inches or so is about the minimum that will give satisfactory results. Even then the pointer has to be arranged in such a fashion that there is sufficient leverage to magnify the very small expansion of the tubing.

Drawings 2 and 3 of Fig. 98 show two constructions of simple bimetallic thermometers; 2 will need the metal of higher coefficient expansion on the outside (iron, steel). Two thin lamina of these metals should be riveted, welded, or soldered together.

Rolling is most desirable if there is equipment for that purpose on hand. Normally, soldering will satisfy minimum requirements. A thermometer constructed of two soldered strips of metal will not stand up for very long, because solder has a tendency to break loose under stress. Instead of the ring form shown in 2, a straight form indicated in 3 may be used for the bimetallic thermometer. This type of instrument is, in its home-made form, rather insensitive.

ESTIMATION OF CLOUD HEIGHT

There is no reliable way to determine exact cloud height by estimation. However, when an observer is required to give his best estimate without the aid of cloud measuring devices, he may resort to one or more methods. Comparison of cloud height against a mountain or hill may give a direct answer if the height of the land is known and if the cloud touches the mountain side or is directly over the crest.

For clouds caused by convection, such as cumulus, and strato-cumulus during high winds, the temperature and dew-point formula may be used to obtain an indication of the level where condensation takes place. The approximate formula is

$$\text{Height of cloud} = \frac{T - \text{D.P.}}{4.5} \times 1,000 \text{ feet}$$

where T and D.P. are in degrees Fahrenheit. For example, if the temperature is 65° F. and the dew point is 56° F., the difference is 9° F. and

$$\text{Height of convective cloud} = \frac{9}{4.5} \times 1,000 = 2,000 \text{ feet}$$

Estimates based on experience and cloud type may serve when the foregoing methods do not apply.

REFERENCES

1. BYERS, H. R. *General Meteorology*. McGraw-Hill Book Co., New York, 1944.
2. HUMPHREYS, W. J. *Physics of the Air*. Third Edition. McGraw-Hill Book Co., New York, 1940.
3. MIDDLETON, W. E. KNOWLES, *Meteorological Instruments*. The University of Toronto Press, Toronto, Canada, 1942.
4. MIDDLETON, W. E. KNOWLES, *Visibility in Meteorology*. University of Toronto Press, Toronto, Canada, 1935.
5. PETTERSSEN, SVERRE. *Introduction to Meteorology*. McGraw-Hill Book Co., New York, 1941.
6. *Smithsonian Meteorological Tables*, Smithsonian Institution, Washington, D. C., 1931.
7. TAYLOR, GEORGE F. *Aeronautical Meteorology*. Pitman Publishing Corp., New York, 1938.
8. *Barometers and the Measurement of Atmospheric Pressure*. Seventh Edition, U. S. Department of Commerce, Weather Bureau. U. S. Government Printing Office.
9. *Codes for Cloud Forms and States of the Sky*. Circular S. U. S. Department of Commerce, Weather Bureau. U. S. Government Printing Office.
10. *Instructions for Airway Meteorological Service*. Fifth Edition, 1941. U. S. Department of Commerce, Weather Bureau. U. S. Government Printing Office.
11. *Instructions for Cooperative Observers*. Circulars B and C, Ninth Edition. U. S. Department of Commerce, Weather Bureau. U. S. Government Printing Office.
12. *Instructions for Erecting and Using Weather Bureau Nephoscope, 1919 Pattern*. Circular I. U. S. Department of Commerce, Weather Bureau. U. S. Government Printing Office.
13. *Instructions for Making Pilot Balloon Observations (W. B. No. 1278)*. Circular O. U. S. Department of Commerce, Weather Bureau. U. S. Government Printing Office.
14. *Instructions to Marine Meteorological Observers*, Seventh Edition, Circular M. U. S. Department of Commerce, Weather Bureau. U. S. Government Printing Office.
15. *Instructions for Modulated Audio Frequency Radiosonde Observations*, Fifth Edition, 1945. U. S. Department of Commerce, Weather Bureau. U. S. Government Printing Office.

16. *International Atlas of Clouds and of States of the Sky.* À l'Office National Météorologique, Paris, 1932.
17. *Measurement of Precipitation*, Circular E, Fourth Edition. U. S. Department of Commerce, Weather Bureau. U. S. Government Printing Office.
18. *Meteorological Observer's Handbook*, 1942. H. M. Stationery Office, London.
19. HAYNES, B. C. *Meteorology for Pilots.* Civil Aeronautics Bul. 25. U. S. Department of Commerce, CAA, Washington, D. C. U. S. Government Printing Office.
20. *Psychometric Tables for Obtaining the Vapor Pressure, Relative Humidity, and Temperature of the Dew Point.* W. B. No. 235, U. S. Department of Commerce, Weather Bureau. U. S. Government Printing Office.
21. *The Elements of Surface Weather Observation Correspondence Course* (Mimeograph). U. S. Department of Commerce, Weather Bureau.

BIBLIOGRAPHY

Clouds

ABBE, CLEVELAND. "Espy's Nepheloscope." *School Science and Mathematics*, Vol. 7, 1907, pp. 586–587.

ABBE, CLEVELAND. "Espy's Nepheloscope." *U. S. Weather Bureau, Monthly Weather Review*, Vol. 35, 1907, p. 123.

BESSON, LOUIS. "A New Nephoscope." *U. S. Weather Bureau, Monthly Weather Review*, Vol. 32, 1904, pp. 13–14.

BESSON, LOUIS. "A Vertical Component of the Movement of Clouds Measured by the Nephoscope." *U. S. Weather Bureau, Monthly Weather Review*, Vol. 31, 1903, pp. 22–24.

BROOKS, CHARLES F. "A Celestial Searchlight." *Science*, New York, Vol. 72, Sept. 5, 1930, p. 244.

BROOKS, CHARLES F. "Cloud Transformations." *Meteorological Magazine*, London, Vol. 57, December, 1922, pp. 303–304.

BROOKS, CHARLES F. "Small Balloons for 'Laboratory' Instruction in Meteorology." *American Meteorological Soc. Bulletin*, Worcester, Mass., Vol. 10, February, 1929, pp. 39–40.

DIGHT, F. H. "The Significance of Nephoscope Observations." *Meteorological Magazine*, London, Vol. 65, 1931, pp. 280–284.

DINES, L. H. G. "Humidity Observations as an Aid to Estimating Cloud-Height." *Meteorological Magazine*, London, Vol. 56, September, 1921, pp. 226–228.

ESPY, J. P. "Nepheloscope." *American Philosophical Society, Proceedings*, Vol. 2, 1841, pp. 128–130.

KADEL, B. C. "Instructions for Erecting and Using Weather Bureau Nephoscope, 1919 Pattern." Washington, 1936, 11 pp. *U. S. Weather Bureau Circular I.*

KIDSON, E. "Cloud-Heights from Melbourne Observatory Photographs." Wellington, 1923, pp. 153–192.

MARVIN, C. F. "Cloud Observations and an Improved Nephoscope." *U. S. Weather Bureau, Monthly Weather Review*, Vol. 24, 1896, pp. 9–13.

MIDDLETON, W. E. K. "Instruments for Investigating Clouds." *Meteorological Instruments*, Toronto, 1942, pp. 166–178.

MIDDLETON, W. E. K. "On the Theory of the Ceiling Projector." *Journal of the Optical Society of America*, Vol. 29, 1939, pp. 340–349.

THOMSON, ANDREW. "Clouds in the Stratosphere." *Science*, New York, Vol. 77, Jan. 27, 1933, p. 9.

THOMSON, ANDREW. "Mother of Pearl Clouds." *Roy. Astron. Socy. Can., Journal*, Toronto, Vol. 26, December, 1932, pp. 437–441.

U. S. WEATHER BUREAU. *Cloud Height Recorder.* Washington, 1943, 3 pp.
U. S. WEATHER BUREAU. *Modulated Light Projector (air-cooled type).* Washington, 1943, 5 pp.

VISIBILITY

ABNEY, W. DE W., and WATSON, W. "The Threshold of Vision for Different Colored Lights." *Royal Society of London, Philosophical Transactions* A, Vol. 216, 1915, pp. 91–128.
ANDRUS, C. G., and others. "Ceiling and Visibility in U. S." *U. S. Weather Bureau, Monthly Weather Review,* Vol. 58, 1930, pp. 198–204.
BENNETT, M. G. "The Visual Range of Lights at Night and its Relation to the Visual Range of Ordinary Objects by Day." *Royal Meteorology Society Quarterly Journal,* Vol. 58, 1932, pp. 259–271.
BENNETT, M. G. "Further Conclusions Concerning Visibility by Day and Night." *Royal Meteorological Society, Quarterly Journal,* Vol. 61, 1935, pp. 179–188.
COBB, P. W. "Dark Adaptation, with Especial Reference to the Problems of Night-Flying." *Psychological Review,* Vol. 26, 1919, pp. 428–453.
CRITTENDEN, ARTHUR H. "The Measurement of Light." *Washington Academy of Science, Journal,* Vol. 13, 1923, pp. 69–90.
COMPTON, ARTHUR H. "What is Light?" *Annual Report, 1929, Smithsonian Institution, Publ.* 3034, pp. 215–228.
DINES and GATTY. "Visibility." *Meteorological Magazine,* Vol. 56, No. 669, pp. 250–252.
GEORGE, J. J. *The Causes and Forecasting of Low Ceilings and Fogs at New Orleans Airport.* Atlanta, Ga., Eastern Airlines, Inc., 1940.
GOLD, E. "Range of Visibility in a Fog." *Meteorological Magazine,* Vol. 65, 1930, pp. 11–12.
HOUGHTON, H. G. "On Relation between Visibility and the Constitution of Clouds and Fog." *Journal of Aeronautical Sciences,* Vol. 6, 1939, pp. 408–411.
HOUGHTON, H. G. "Size and Distribution of Fog Particles." *Physics,* Vol. 2, 1932, pp. 467–475.
HOUGHTON, H. G. "Transmission of Visible Light through Fog." *Physical Review,* Vol. 38, 1931, pp. 152–158.
HOUGHTON, H. G. "On Relation between Visibility and the Constitution of Clouds and Fog." *Journal of Aeronautical Sciences,* Vol. 6, 1939, pp. 408–411.
JEFFREYS, HAROLD. "The Shape of the Sky." *Meteorological Magazine,* 1921, No. 667, pp. 173–177.
MIDDLETON, W. E. K. "Apparent Color of Lights at Night with an Observation of 'Blue Fog.'" *Royal Meteorological Society, Quarterly Journal,* Vol. 62, 1936, pp. 473–480.
MIDDLETON, W. E. K. "How Far Can I See?" *Scientific Monthly,* Vol. 41, 1935, pp. 343–346.
MIDDLETON, W. E. K. "Measurement of Visibility at Night." *Royal Society of Canada, Transactions,* Sec. 3, Vol. 25, 1931, pp. 39–48.

MIDDLETON, W. E. K. "On Colors of Distant Objects and the Visual Range of Colored Objects." *Royal Society of Canada, Transactions*, Sec. 3, Vol. 29, 1935, pp. 122–154.

PICK, W. H. "A Note of Inter-Relation between Visibility and Relative Humidity." *Royal Meteorological Society, Quarterly Journal*, Vol. 56, 1930, pp. 183–184.

PICK, W. H. "Ground Horizontal Visibility and Convection." *Meteorological Magazine*, Vol. 62, 1927, pp. 289–290.

PICK, W. H. "Surface Wind and Horizontal Visibility." *Meteorological Magazine*, Vol. 63, 1928, pp. 114–116.

STRATTON, J. A., and HOUGHTON, H. G. "Theoretical Investigation of the Transmission of Light through Fog." *Physical Review*, Vol. 38, 1931, pp. 159–165.

TEMPERATURE

AITKEN, JOHN. "Thermometer Screens." *Royal Society of Edinburgh, Proceedings*, Vol. 40, pt. 2, sess. 1919–1920, pp. 172–181.

ANDSON, W. "Temperature of Air and Rivers." Note on paper by W. Andson. *Symons's Meteorological Magazine*, London, Vol. 38, pp. 4–6.

BROOKS, CHARLES F. "Unreal 'Errors' of Thermometers." *Tycos-Rochester*, Rochester, N. Y., Vol. 13, October, 1923, p. 6.

BROOKS, CHARLES F. "Temperatures Outside vs. those inside a Thermometer Shelter." *American Meteorological Society, Bulletin*, Worcester, Mass., Vol. 3, March, 1922, p. 37.

BROWN, S. LEROY. "A New Form of Resistance Thermometer." *Physical Review*, Lancaster, Vol. 5, February, 1915, pp. 126–134.

BROWNELL, BAKER. "A Home-made Air Thermometer." *Scientific American*, New York, Aug. 17, 1907, p. 118.

CHREE, C. "The Effect of Pressure on the Readings of Thermometers." *Royal Meteorological Society, Quarterly Journal*, London, Vol. 53, October, 1927, p. 438.

HARPER, D. R., 3D. "Thermometric Lag." *Bulletin Bureau Standards*, Washington, Vol. 8, March 1, 1913, pp. 659–714.

LANG, H. R. "The Construction of Platinum Thermometers." *Journal of Scientific Instruments*, London, Vol. 2, April, 1925, pp. 228–233.

LASKOWSKI, B. R. "Ground Temperatures vs. Roof Temperatures." *Tycos-Rochester*, Rochester, N. Y., Vol. 17, April, 1927, pp. 68–70.

McCAW, G. T. "The Observation of Air Temperature in the Tropics." *Geographical Journal*, London, Vol. 34, September, 1909, pp. 298–300.

THIESSEN, ALFRED H. "Story of the Thermometer and Its Uses in Agriculture." *Yearbook, U. S. Department of Agriculture*, Washington, 1914, pp. 157–166.

HUMIDITY

GOODMAN, WILLIAM. *Air Conditioning Analysis with Psychrometric Charts and Tables.* New York, The Macmillan Company, 1943.

SHAW, A. N. "Relative Humidity." *Transactions of the Royal Society of Canada*, Ottawa, 1917, pp. 121–127.

Talman, C. G. "What Humidity Is; How It Is Caused." *New York Times,* New York, Aug. 9, 1931, Sec. 9, p. 5.

Torok, Elmer. *Psychrometric Notes and Tables; a Handbook on Psychrometric Principles, Tables and Calculations for Textile Manufacturers, Engineers and Students.* Revised edition. North American Rayon Corporation, New York, 1941.

Windsor, E. V. "Hygrometric Determinations." *Nature,* London, Vol. 67, pp. 463–464.

Wind

Chatley, Herbert, *The Force of the Wind.* London, C. Griffin & Co., Ltd., 1909.

Claudy, C. H. "Wind Velocity and Direction." Extract, *American Inventor,* Washington, Feb. 15, 1902, Vol. 8, No. 13, p. 1.

Dyck, Homer D. "Comparison of Extreme Gust Velocities as Recorded by the Dines Anemometer and 5-Minute Velocities as Recorded by the Robinson Anemometer." *U. S. Weather Bureau, Monthly Weather Review,* Vol. 69, 1941, pp. 301–302.

Dines, William Henry. "Anemometer Comparisons." *Quarterly Journal of the Royal Meteorological Society,* Vol. 18, No. 83, July, 1892, pp. 165–183.

Friez, Julien P. & Sons. *Wind News.* Baltimore, Md.

Rossby, Carl-Gustaf, and R. B. Montgomery. "The Layer of Frictional Influence in Wind and Ocean Currents." *Papers in Physical Oceanography and Meteorology,* Cambridge, Mass., April, 1935, Vol. 3, No. 3.

Spilhaus, Athelstan F. "Analysis of the Cup Anemometer." *Massachusetts Institute of Technology, Professional Notes,* No. 7, Cambridge, Mass., 1934.

Talman, C. F. "The Uses of Weather Vanes." *New York Times Magazine,* New York, June 19, 1932, p. 16.

Whipple, Francis John Welsh. "Notes on the Robinson Anemometer." *Advisory Committee for Aeronautics, Reports and Memoranda,* No. 669, London, 1920.

Pressure

Abbe, Cleveland. "The Barometer as Used at Sea." *U. S. Weather Bureau, Monthly Weather Review,* Vol. 29, 1901, pp. 459–460.

Abbe, Cleveland. "Treatise on Meteorological Apparatus and Methods." *Annual Report of the Chief Signal Officer,* Washington, Appendix 46, Part 2, 1888, 392 pp.

Abercromby, Ralph. "An Improvement in Aneroid Barometers." *Royal Meteorological Society, Quarterly Journal,* Vol. 3, 1877, pp. 87–89.

American Paulin System, Inc. *Origin and Development of the Barometer and Altimeter.* Los Angeles, c1929.

Bendix Aviation Corporation. "Manual for Mercurial Barometer." Bendix, N. J., c1942.

Bigelow, F. H. "Report on the Barometry of the United States, Canada, and the West Indies." *U. S. Weather Bureau, Report of the Chief, 1900–01,* Washington, Vol. 2, 1902.

DINES, L. H. G. "The Dines Float Barograph." *Royal Meteorological Society, Quarterly Journal*, Vol. 55, 1929, pp. 37–53.

FERGUSSON, S. P. "An International Comparison of Standard Barometers." *U. S. Weather Bureau, Monthly Weather Review*, Vol. 62, 1934, pp. 364–366.

FIELD, R. H. *Aneroid Barometer and Altimeter, Their Characteristics and Use in Mapping*. With an appendix, The Field Use of the Aneroid Barometer, by G. C. COOPER. Ottawa, 1931, 36 pp.

FRIEZ, JULIEN P. & SONS. *Friez Barographs*. Baltimore, Md., 1936.

GIBLETT, M. A. "The Effect of the Rolling of a Ship on the Readings of a Marine Mercury Barometer." *Philosophical Magazine*, London, Vol. 46, 1923, pp. 707–716.

HARRISON, E. P. "On Cleaning and Refilling Various Types of Barometers, Together with a Description of Several Usual Patterns." India, *Meteorological Department Memoirs*, Vol. 23, Part 5, 1922.

HUMPHREYS, W. J. "Why the Readings of the Mercurial Barometer Are Corrected for Both Temperature and Latitude and the Readings of the Aneroid Barometer Left Unchanged." *U. S. Weather Bureau, Monthly Weather Review*, Vol. 59, 1931, p. 239.

KELLY, R. D. *Flight Test Procedure Using the Paulin Altimeters and Barometers*. Los Angeles, c1929, 14 pp.

MARVIN, C. F. "Aneroid Barometers." *U. S. Weather Bureau, Monthly Weather Review*, Vol. 26, 1898, pp. 410–412.

MARVIN, C. F. "Barometers and the Measurement of Atmospheric Pressure." Washington, 1941, 91 pp. *U. S. Weather Bureau, Circular F*.

MARVIN, C. F. "A Mercurial Barograph of High Precision." *U. S. Weather Bureau, Monthly Weather Review*, Vol. 36, 1908, pp. 307–313.

PASCAL, BLAISE. *Physical Treatises; the Equilibrium of Liquids and the Weight of the Mass of Air;* tr. by I. H. B. and A. G. H. SPIERS, with introduction and notes by FREDERICK BARRY. Columbia University Press, New York, 1937, 181 pp.

PRECIPITATION

DINES, W. H. "The Tilting Rain-Gauge: A New Autographic Instrument." *Meteorological Magazine*, Vol. 55, 1920, pp. 112–113.

FERGUSSON, S. P. *Self-Recording Rain or Snow Gage*. U. S. Pat. 417357, Dec. 17, 1889, 4 pp.

FRIEZ, J. P. & SONS. *Erection Instructions for Rain Gage Steel Tower Supports (1) for Friez Tube Type Rain Gages, (2) for Friez Recording Rain Gages*. Baltimore, Md., 50 pp.

GOLD, ERNEST. "Exposure of Rain Gauges." *Meteorological Magazine*, Vol. 57, 1922, pp. 231–235.

HORTON, R. E. "The Measurement of Rainfall and Snow." Reprinted from *New England Water Works Association Journal*, Vol. 33, No. 1, 1919, pp. 14–71.

HUDLESTON, F. "On Certain Experiments with Rain Gauge Shields, made during the winter of 1926–27 at Hutton John, Cumberland, in the northeast corner of the Lake District. *British Rainfall*, 1926, pp. 285–293.

KADEL, B. C. "Measurement of Precipitation. Instructions for Measurement and Registration of Precipitation by Means of the Standard Instruments of the U. S. Weather Bureau." *U. S. Weather Bureau Circular E*, Washington, 1936, 25 pp.

LANDSBERG, HELMUT. "A Convenient Heated Precipitation Gage." *American Meteorological Society, Bulletin*, Vol. 20, No. 9, pp. 383–385.

LONG, T. L. "A Comparison of Snowfall Catch in Shielded and Unshielded Precipitation Gauges." Manuscript. April 25, 1941, 5 pp.

MALTAIS, J. B. *A New Recording Rain-Gauge.* Ottawa, 1936, pp. 495–498.

MEARS, J. W. *The Experimental Development of an Automatic Integrating "Intensity" Rain-Gauge without Clockwork.* Institution of Civil Engineers, London, 1923, 29 pp.

RADIOSONDES

BENDIX AVIATION CORPORATION. Friez Instrument Division. *Instructions for Radio-Sonde Equipment.* Baltimore, Md., 1943, 71 pp.

CLARKE, E. T., and KORFF, S. A. "The Radiosonde: the Stratosphere Laboratory." *Journal of the Franklin Institute*, Vol. 232, No. 3, 1941, pp. 217–238, 339–355.

CURTISS, L. F., and others. "An Improved Radio Meteorograph on the Olland Principle." *U. S. National Bureau of Standards, Journal of Research*, Vol. 22, 1939, pp. 97–103.

DIAMOND, HARRY. "Comparisons of Soundings with Radio Meteorographs, Aerographs and Meteorographs." *American Meteorological Society, Bulletin*, Vol. 19, 1938, pp. 129–141.

DIAMOND, HARRY, and others. "Development of a Radio Meteorograph System for the Navy Department." *American Meteorological Society, Bulletin*, Vol. 18, 1937, pp. 73–99.

DIAMOND, HARRY, and others. "An Improved Radio Sonde and Its Performance," *U. S. National Bureau of Standards, Journal of Research*, Vol. 25, 1940, pp. 327–367.

DIAMOND, HARRY. "Recent Application of Radio to the Remote Indication of Meteorological Elements." *Electrical Engineering*, Vol. 59, 1941, pp. 136–167.

DUNMORE, F. W. "An Electric Hygrometer and Its Application to Radio Meteorography." *U. S. National Bureau of Standards, Journal of Research*, Vol. 20, 1935, pp. 723–744.

DUNMORE, F. W. "An Improved Electric Hygrometer." *U. S. National Bureau of Standards, Journal of Research*, Vol. 23, 1939, pp. 701–714.

FRIEZ, JULIEN P. & SONS. *Friez Ray-Sonde, Diamond-Hinman System.* Baltimore, Md., 1938, 20 pp.

FRIEZ, JULIEN P. & SONS. *Instructions for Installation, Operation, and Maintenance of the Complete Friez Ray-Sonde, Diamond-Hinman System.* Baltimore, Md., 1939, 49 pp.

HAFER, L. F. "Comparative Observations with Friez-type Radiosondes and Fergusson Meteorographs." *U. S. Weather Bureau, Monthly Weather Review*, Vol. 70, 1942, pp. 203–208.

HARMANTAS, C. "Upper-air Temperatures Obtained by Use of Radiosonde," *Temperature, Its Measurement and Control,* American Institute of Physics, New York, 1941, pp. 381–388.

LANGE, K. O. "Radio-Meteorographs." *American Meteorological Society, Bulletin,* Vol. 16, 1935, pp. 233–236, 267–271, 297–300.

LITTLE, D. M. "Contributions to the Development of the Radio-Meteorograph by the United States Weather Bureau." *American Geophysical Union, Transactions,* 1937, pp. 138–141.

MAIER, O. C., and WOOD, L. E. "The Galcit Radio Meteorograph." *Journal of the Aeronautical Sciences,* Vol. 4, 1937, No. 10.

APPENDIX

TABLE A–1

Barometric Inches (Mercury) into Millibars

1 inch = 33.86395 mb.; 1 mb. = 0.02952993 inch

Inches	.00	.01	.02	.03	.04	.05	.06	.07	.08	.09
	mb.	mb.	mb.	mb.	mb.	mb.	mb.	mb.	mb.	mb.
26.0	880.5	880.8	881.1	881.5	881.8	882.2	882.5	882.8	883.2	883.5
26.1	883.8	884.2	884.5	884.9	885.2	885.5	885.9	886.2	886.6	886.9
26.2	887.2	887.6	887.9	888.3	888.6	888.9	889.3	889.6	889.9	890.3
26.3	890.6	891.0	891.3	891.6	892.0	892.3	892.7	893.0	893.3	893.7
26.4	894.0	894.3	894.7	895.0	895.4	895.7	896.0	896.4	896.7	897.1
26.5	897.4	897.7	898.1	898.4	898.7	899.1	899.4	899.8	900.1	900.4
26.6	900.8	901.1	901.5	901.8	902.1	902.5	902.8	903.2	903.5	903.8
26.7	904.2	904.5	904.8	905.2	905.5	905.9	906.2	906.5	906.9	907.2
26.8	907.6	907.9	908.2	908.6	908.9	909.2	909.6	909.9	910.3	910.6
26.9	910.9	911.3	911.6	912.0	912.3	912.6	913.0	913.3	913.6	914.0
27.0	914.3	914.7	915.0	915.3	915.7	916.0	916.4	916.7	917.0	917.4
27.1	917.7	918.1	918.4	918.7	919.1	919.4	919.7	920.1	920.4	920.8
27.2	921.1	921.4	921.8	922.1	922.5	922.8	923.1	923.5	923.8	924.1
27.3	924.5	924.8	925.2	925.5	925.8	926.2	926.5	926.9	927.2	927.5
27.4	927.9	928.2	928.5	928.9	929.2	929.6	929.9	930.2	930.6	930.9
27.5	931.3	931.6	931.9	932.3	932.6	933.0	933.3	933.6	934.0	934.3
27.6	934.6	935.0	935.3	935.7	936.0	936.3	936.7	937.0	937.4	937.7
27.7	938.0	938.4	938.7	939.0	939.4	939.7	940.1	940.4	940.7	941.1
27.8	941.4	941.8	942.1	942.4	942.8	943.1	943.4	943.8	944.1	944.5
27.9	944.8	945.1	945.5	945.8	946.2	946.5	946.8	947.2	947.5	947.9
28.0	948.2	948.5	948.9	949.2	949.5	949.9	950.2	950.6	950.9	951.2
28.1	951.6	951.9	952.3	952.6	952.9	953.3	953.6	953.9	954.3	954.6
28.2	955.0	955.3	955.6	956.0	956.3	956.7	957.0	957.3	957.7	958.0
28.3	958.3	958.7	959.0	959.4	959.7	960.0	960.4	960.7	961.1	961.4
28.4	961.7	962.1	962.4	962.8	963.1	963.4	963.8	964.1	964.4	964.8
28.5	965.1	965.5	965.8	966.1	966.5	966.8	967.2	967.5	967.8	968.2
28.6	968.5	968.8	969.2	969.5	969.9	970.2	970.5	970.9	971.2	971.6
28.7	971.9	972.2	972.6	972.9	973.2	973.6	973.9	974.3	974.6	974.9
28.8	975.3	975.6	976.0	976.3	976.6	977.0	977.3	977.7	978.0	978.3
28.9	978.7	979.0	979.3	979.7	980.0	980.4	980.7	981.0	981.4	981.7
29.0	982.1	982.4	982.7	983.1	983.4	983.7	984.1	984.4	984.8	985.1
29.1	985.4	985.8	986.1	986.5	986.8	987.1	987.5	987.8	988.2	988.5
29.2	988.8	989.2	989.5	989.8	990.2	990.5	990.9	991.2	991.5	991.9
29.3	992.2	992.6	992.9	993.2	993.6	993.9	994.2	994.6	994.9	995.3
29.4	995.6	995.9	996.3	996.6	997.0	997.3	997.6	998.0	998.3	998.6
29.5	999.0	999.3	999.7	1000.0	1000.3	1000.7	1001.0	1001.4	1001.7	1002.0
29.6	1002.4	1002.7	1003.1	1003.4	1003.7	1004.1	1004.4	1004.7	1005.1	1005.4
29.7	1005.8	1006.1	1006.4	1006.8	1007.1	1007.5	1007.8	1008.1	1008.5	1008.8
29.8	1009.1	1009.5	1009.8	1010.2	1010.5	1010.8	1011.2	1011.5	1011.9	1012.2
29.9	1012.5	1012.9	1013.2	1013.5	1013.9	1014.2	1014.6	1014.9	1015.2	1015.6
30.0	1015.9	1016.3	1016.6	1016.9	1017.3	1017.6	1018.0	1018.3	1018.6	1019.0
30.1	1019.3	1019.6	1020.0	1020.3	1020.7	1021.0	1021.3	1021.7	1022.0	1022.4
30.2	1022.7	1023.0	1023.4	1023.7	1024.0	1024.4	1024.7	1025.1	1025.4	1025.7
30.3	1026.1	1026.4	1026.8	1027.1	1027.4	1027.8	1028.1	1028.4	1028.8	1029.1
30.4	1029.5	1029.8	1030.1	1030.5	1030.8	1031.2	1031.5	1031.8	1032.2	1032.5
30.5	1032.9	1033.2	1033.5	1033.9	1034.2	1034.5	1034.9	1035.2	1035.6	1035.9
30.6	1036.2	1036.6	1036.9	1037.3	1037.6	1038.0	1038.3	1038.6	1038.9	1039.3
30.7	1039.6	1040.0	1040.3	1040.6	1041.0	1041.3	1041.7	1042.0	1042.3	1042.7
30.8	1043.0	1043.3	1043.7	1044.0	1044.4	1044.7	1045.0	1045.4	1045.7	1046.1
30.9	1046.4	1046.7	1047.1	1047.4	1047.8	1048.1	1048.4	1048.8	1049.1	1049.5
31.0	1049.8	1050.1	1050.5	1050.8	1051.1	1051.5	1051.8	1052.2	1052.5	1052.8
31.1	1053.2	1053.5	1053.8	1054.2	1054.5	1054.9	1055.2	1055.5	1055.9	1056.2
31.2	1056.6	1056.9	1057.2	1057.6	1057.9	1058.2	1058.6	1058.9	1059.3	1059.6
31.3	1059.9	1060.3	1060.6	1061.0	1061.3	1061.6	1062.0	1062.3	1062.7	1063.0
31.4	1063.3	1063.7	1064.0	1064.3	1064.7	1065.0	1065.4	1065.7	1066.0	1066.4
31.5	1066.7	1067.1	1067.4	1067.7	1068.1	1068.4	1068.7	1069.1	1069.4	1069.8
31.6	1070.1	1070.4	1070.8	1071.1	1071.5	1071.8	1072.1	1072.5	1072.8	1073.1
31.7	1073.5	1073.8	1074.2	1074.5	1074.8	1075.2	1075.5	1075.9	1076.2	1076.5
31.8	1076.9	1077.2	1077.6	1077.9	1078.2	1078.6	1078.9	1079.2	1079.6	1079.9
31.9	1080.3	1080.6	1080.9	1081.3	1081.6	1082.0	1082.3	1082.6	1083.0	1083.3

TABLE A–2
Fahrenheit to Centigrade Temperatures

°F.	.0	.1	.2	.3	.4	.5	.6	.7	.8	.9
	°C.	°C.	°C.	°C.	°C.	°C.	°C.	°C.	°C.	°C.
− 0	−17.78	−17.83	−17.89	−17.94	−18.00	−18.06	−18.11	−18.17	−18.22	−18.28
1	18.33	18.39	18.44	18.50	18.56	18.61	18.67	18.72	18.78	18.83
2	18.89	18.94	19.00	19.06	19.11	19.17	19.22	19.28	19.33	19.39
3	19.44	19.50	19.56	19.61	19.67	19.72	19.78	19.83	19.89	19.94
4	20.00	20.06	20.11	20.17	20.22	20.28	20.33	20.39	20.44	20.50
− 5	−20.56	−20.61	−20.67	−20.72	−20.78	−20.83	−20.89	−20.94	−21.00	−21.06
6	21.11	21.17	21.22	21.28	21.33	21.39	21.44	21.50	21.56	21.61
7	21.67	21.72	21.78	21.83	21.89	21.94	22.00	22.06	22.11	22.17
8	22.22	22.28	22.33	22.39	22.44	22.50	22.56	22.61	22.67	22.72
9	22.78	22.83	22.89	22.94	23.00	23.06	23.11	23.17	23.22	23.28
−10	−23.33	−23.39	−23.44	−23.50	−23.56	−23.61	−23.67	−23.72	−23.78	−23.83
11	23.89	23.94	24.00	24.06	24.11	24.17	24.22	24.28	24.33	24.39
12	24.44	24.50	24.56	24.61	24.67	24.72	24.78	24.83	24.89	24.94
13	25.00	25.06	25.11	25.17	25.22	25.28	25.33	25.39	25.44	25.50
14	25.56	25.61	25.67	25.72	25.78	25.83	25.89	25.94	26.00	26.06
−15	−26.11	−26.17	−26.22	−26.28	−26.33	−26.39	−26.44	−26.50	−26.56	−26.61
16	26.67	26.72	26.78	26.83	26.89	26.94	27.00	27.06	27.11	27.17
17	27.22	27.28	27.33	27.39	27.44	27.50	27.56	27.61	27.67	27.72
18	27.78	27.83	27.89	27.94	28.00	28.06	28.11	28.17	28.22	28.28
19	28.33	28.39	28.44	28.50	28.56	28.61	28.67	28.72	28.78	28.83
−20	−28.89	−28.94	−29.00	−29.06	−29.11	−29.17	−29.22	−29.28	−29.33	−29.39
21	29.44	29.50	29.56	29.61	29.67	29.72	29.78	29.83	29.89	29.94
22	30.00	30.06	30.11	30.17	30.22	30.28	30.33	30.39	30.44	30.50
23	30.56	30.61	30.67	30.72	30.78	30.83	30.89	30.94	31.00	31.06
24	31.11	31.17	31.22	31.28	31.33	31.39	31.44	31.50	31.56	31.61
−25	−31.67	−31.72	−31.78	−31.83	−31.89	−31.94	−32.00	−32.06	−32.11	−32.17
26	32.22	32.28	32.33	32.39	32.44	32.50	32.56	32.61	32.67	32.72
27	32.78	32.83	32.89	32.94	33.00	33.06	33.11	33.17	33.22	33.28
28	33.33	33.39	33.44	33.50	33.56	33.61	33.67	33.72	33.78	33.83
29	33.89	33.94	34.00	34.06	34.11	34.17	34.22	34.28	34.33	34.39
−30	−34.44	−34.50	−34.56	−34.61	−34.67	−34.72	−34.78	−34.83	−34.89	−34.94
31	35.00	35.06	35.11	35.17	35.22	35.28	35.33	35.39	35.44	35.50
32	35.56	35.61	35.67	35.72	35.78	35.83	35.89	35.94	36.00	36.06
33	36.11	36.17	36.22	36.28	36.33	36.39	36.44	36.50	36.56	36.61
34	36.67	36.72	36.78	36.83	36.89	36.94	37.00	37.06	37.11	37.17
−35	−37.22	−37.28	−37.33	−37.39	−37.44	−37.50	−37.56	−37.61	−37.67	−37.72
36	37.78	37.83	37.89	37.94	38.00	38.06	38.11	38.17	38.22	38.28
37	38.33	38.39	38.44	38.50	38.56	38.61	38.67	38.72	38.78	38.83
38	38.89	38.94	39.00	39.06	39.11	39.17	39.22	39.28	39.33	39.39
39	39.44	39.50	39.56	39.61	39.67	39.72	39.78	39.83	39.89	39.94
−40	−40.00	−40.06	−40.11	−40.17	−40.22	−40.28	−40.33	−40.39	−40.44	−40.50
41	40.56	40.61	40.67	40.72	40.78	40.83	40.89	40.94	41.00	41.06
42	41.11	41.17	41.22	41.28	41.33	41.39	41.44	41.50	41.56	41.61
43	41.67	41.72	41.78	41.83	41.89	41.94	42.00	42.06	42.11	42.17
44	42.22	42.28	42.33	42.39	42.44	42.50	42.56	42.61	42.67	42.72
−45	−42.78	−42.83	−42.89	−42.94	−43.00	−43.06	−43.11	−43.17	−43.22	−43.28
46	43.33	43.39	43.44	43.50	43.56	43.61	43.67	43.72	43.78	43.83
47	43.89	43.94	44.00	44.06	44.11	44.17	44.22	44.28	44.33	44.39
48	44.44	44.50	44.56	44.61	44.67	44.72	44.78	44.83	44.89	44.94
49	45.00	45.06	45.11	45.17	45.22	45.28	45.33	45.39	45.44	45.50
−50	−45.56	−45.61	−45.67	−45.72	−45.78	−45.83	−45.89	−45.94	−46.00	−46.06
51	46.11	46.17	46.22	46.28	46.33	46.39	46.44	46.50	46.56	46.61
52	46.67	46.72	46.78	46.83	46.89	46.94	47.00	47.06	47.11	47.17
53	47.22	47.28	47.33	47.39	47.44	47.50	47.56	47.61	47.67	47.72
54	47.78	47.83	47.89	47.94	48.00	48.06	48.11	48.17	48.22	48.28
−55	−48.33	−48.39	−48.44	−48.50	−48.56	−48.61	−48.67	−48.72	−48.78	−48.83
56	48.89	48.94	49.00	49.06	49.11	49.17	49.22	49.28	49.33	49.39
57	49.44	49.50	49.56	49.61	49.67	49.72	49.78	49.83	49.89	49.94
58	50.00	50.06	50.11	50.17	50.22	50.28	50.33	50.39	50.44	50.50
59	50.56	50.61	50.67	50.72	50.78	50.83	50.89	50.94	51.00	51.06

Fahrenheit to Centigrade Temperatures

°F.	.0	.1	.2	.3	.4	.5	.6	.7	.8	.9
	°C.	°C.	°C.	°C.	°C.	°C.	°C.	°C.	°C.	°C
+60	+15.56	+15.61	+15.67	+15.72	+15.78	+15.83	+15.89	+15.94	+16.00	+16.06
59	15.00	15.06	15.11	15.17	15.22	15.28	15.33	15.39	15.44	15.50
58	14.44	14.50	14.56	14.61	14.67	14.72	14.78	14.83	14.89	14.94
57	13.89	13.94	14.00	14.06	14.11	14.17	14.22	14.28	14.33	14.39
56	13.33	13.39	13.44	13.50	13.56	13.61	13.67	13.72	13.78	13.83
+55	+12.78	+12.83	+12.89	+12.94	+13.00	+13.06	+13.11	+13.17	+13.22	+13.28
54	12.22	12.28	12.33	12.39	12.44	12.50	12.56	12.61	12.67	12.72
53	11.67	11.72	11.78	11.83	11.89	11.94	12.00	12.06	12.11	12.17
52	11.11	11.17	11.22	11.28	11.33	11.39	11.44	11.50	11.56	11.61
51	10.56	10.61	10.67	10.72	10.78	10.83	10.89	10.94	11.00	11.06
+50	+10.00	+10.06	+10.11	+10.17	+10.22	+10.28	+10.33	+10.39	+10.44	+10.50
49	9.44	9.50	9.56	9.61	9.67	9.72	9.78	9.83	9.89	9.94
48	8.89	8.94	9.00	9.06	9.11	9.17	9.22	9.28	9.33	9.39
47	8.33	8.39	8.44	8.50	8.56	8.61	8.67	8.72	8.78	8.83
46	7.78	7.83	7.89	7.94	8.00	8.06	8.11	8.17	8.22	8.28
+45	+ 7.22	+ 7.28	+ 7.33	+ 7.39	+ 7.44	+ 7.50	+ 7.56	+ 7.61	+ 7.67	+ 7.72
44	6.67	6.72	6.78	6.83	6.89	6.94	7.00	7.06	7.11	7.17
43	6.11	6.17	6.22	6.28	6.33	6.39	6.44	6.50	6.56	6.61
42	5.56	5.61	5.67	5.72	5.78	5.83	5.89	5.94	6.00	6.06
41	5.00	5.06	5.11	5.17	5.22	5.28	5.33	5.39	5.44	5.50
+40	+ 4.44	+ 4.50	+ 4.56	+ 4.61	+ 4.67	+ 4.72	+ 4.78	+ 4.83	+ 4.89	+ 4.94
39	3.89	3.94	4.00	4.06	4.11	4.17	4.22	4.28	4.33	4.39
38	3.33	3.39	3.44	3.50	3.56	3.61	3.67	3.72	3.78	3.83
37	2.78	2.83	2.89	2.94	3.00	3.06	3.11	3.17	3.22	3.28
36	2.22	2.28	2.33	2.39	2.44	2.50	2.56	2.61	2.67	2.72
+35	+ 1.67	+ 1.72	+ 1.78	+ 1.83	+ 1.89	+ 1.94	+ 2.00	+ 2.06	+ 2.11	+ 2.17
34	+ 1.11	+ 1.17	+ 1.22	+ 1.28	+ 1.33	+ 1.39	+ 1.44	+ 1.50	+ 1.56	+ 1.61
33	+ 0.56	+ 0.61	+ 0.67	+ 0.72	+ 0.78	+ 0.83	+ 0.89	+ 0.94	+ 1.00	+ 1.06
32	0.00	+ 0.06	+ 0.11	+ 0.17	+ 0.22	+ 0.28	+ 0.33	+ 0.39	+ 0.44	+ 0.50
31	- 0.56	- 0.50	- 0.44	- 0.39	- 0.33	- 0.28	- 0.22	- 0.17	- 0.11	- 0.06
+30	- 1.11	- 1.06	- 1.00	- 0.94	- 0.89	- 0.83	- 0.78	- 0.72	- 0.67	- 0.61
29	1.67	1.61	1.56	1.50	1.44	1.39	1.33	1.28	1.22	1.17
28	2.22	2.17	2.11	2.06	2.00	1.94	1.89	1.83	1.78	1.72
27	2.78	2.72	2.67	2.61	2.56	2.50	2.44	2.39	2.33	2.28
26	3.33	3.28	3.22	3.17	3.11	3.06	3.00	2.94	2.89	2.83
+25	- 3.89	- 3.83	- 3.78	- 3.72	- 3.67	- 3.61	- 3.56	- 3.50	- 3.44	- 3.39
24	4.44	4.39	4.33	4.28	4.22	4.17	4.11	4.06	4.00	3.94
23	5.00	4.94	4.89	4.83	4.78	4.72	4.67	4.61	4.56	4.50
22	5.56	5.50	5.44	5.39	5.33	5.28	5.22	5.17	5.11	5.06
21	6.11	6.06	6.00	5.94	5.89	5.83	5.78	5.72	5.67	5.61
+20	- 6.67	- 6.61	- 6.56	- 6.50	- 6.44	- 6.39	- 6.33	- 6.28	- 6.22	- 6.17
19	7.22	7.17	7.11	7.06	7.00	6.94	6.89	6.83	6.78	6.72
18	7.78	7.72	7.67	7.61	7.56	7.50	7.44	7.39	7.33	7.28
17	8.33	8.28	8.22	8.17	8.11	8.06	8.00	7.94	7.89	7.83
16	8.89	8.83	8.78	8.72	8.67	8.61	8.56	8.50	8.44	8.39
+15	- 9.44	- 9.39	- 9.33	- 9.28	- 9.22	- 9.17	- 9.11	- 9.06	- 9.00	- 8.94
14	10.00	9.94	9.89	9.83	9.78	9.72	9.67	9.61	9.56	9.50
13	10.56	10.50	10.44	10.39	10.33	10.28	10.22	10.17	10.11	10.06
12	11.11	11.06	11.00	10.94	10.89	10.83	10.78	10.72	10.67	10.61
11	11.67	11.61	11.56	11.50	11.44	11.39	11.33	11.28	11.22	11.17
+10	-12.22	-12.17	-12.11	-12.06	-12.00	-11.94	-11.89	-11.83	-11.78	-11.72
9	12.78	12.72	12.67	12.61	12.56	12.50	12.44	12.39	12.33	12.28
8	13.33	13.28	13.22	13.17	13.11	13.06	13.00	12.94	12.89	12.83
7	13.89	13.83	13.78	13.72	13.67	13.61	13.56	13.50	13.44	13.39
6	14.44	14.39	14.33	14.28	14.22	14.17	14.11	14.06	14.00	13.94
+ 5	-15.00	-14.94	-14.89	-14.83	-14.78	-14.72	-14.67	-14.61	-14.56	-14.50
4	15.56	15.50	15.44	15.39	15.33	15.28	15.22	15.17	15.11	15.06
3	16.11	16.06	16.00	15.94	15.89	15.83	15.78	15.72	15.67	15.61
2	16.67	16.61	16.56	16.50	16.44	16.39	16.33	16.28	16.22	16.17
1	17.22	17.17	17.11	17.06	17.00	16.94	16.89	16.83	16.78	16.72
+ 0	17.78	17.72	17.67	17.61	17.56	17.50	17.44	17.39	17.33	17.28

TABLE A–2 (Continued)
Fahrenheit to Centigrade Temperatures

°F.	.0	.1	.2	.3	.4	.5	.6	.7	.8	.9
	°C.	°C.	°C.	°C.	°C.	°C.	°C.	°C.	°C.	°C.
+120	+48.89	+48.94	+49.00	+49.06	+49.11	+49.17	+49.22	+49.28	+49.33	+49.39
119	48.33	48.39	48.44	48.50	48.56	48.61	48.67	48.72	48.78	48.83
118	47.78	47.83	47.89	47.94	48.00	48.06	48.11	48.17	48.22	48.28
117	47.22	47.28	47.33	47.39	47.44	47.50	47.56	47.61	47.67	47.72
116	46.67	46.72	46.78	46.83	46.89	46.94	47.00	47.06	47.11	47.17
+115	+46.11	+46.17	+46.22	+46.28	+46.33	+46.39	+46.44	+46.50	+46.56	+46.61
114	45.56	45.61	45.67	45.72	45.78	45.83	45.89	45.94	46.00	46.06
113	45.00	45.06	45.11	45.17	45.22	45.28	45.33	45.39	45.44	45.50
112	44.44	44.50	44.56	44.61	44.67	44.72	44.78	44.83	44.89	44.94
111	43.89	43.94	44.00	44.06	44.11	44.17	44.22	44.28	44.33	44.39
+110	+43.33	+43.39	+43.44	+43.50	+43.56	+43.61	+43.67	+43.72	+43.78	+43.83
109	42.78	42.83	42.89	42.94	43.00	43.06	43.11	43.17	43.22	43.28
108	42.22	42.28	42.33	42.39	42.44	42.50	42.56	42.61	42.67	42.72
107	41.67	41.72	41.78	41.83	41.89	41.94	42.00	42.06	42.11	42.17
106	41.11	41.17	41.22	41.28	41.33	41.39	41.44	41.50	41.56	41.61
+105	+40.56	+40.61	+40.67	+40.72	+40.78	+40.83	+40.89	+40.94	+41.00	+41.06
104	40.00	40.06	40.11	40.17	40.22	40.28	40.33	−40.39	40.44	40.50
103	39.44	39.50	39.56	39.61	39.67	39.72	39.78	39.83	39.89	39.94
102	38.89	38.94	39.00	39.06	39.11	39.17	39.22	39.28	39.33	39.39
101	38.33	38.39	38.44	38.50	38.56	38.61	38.67	38.72	38.78	38.83
+100	+37.78	+37.83	+37.89	+37.94	+38.00	+38.06	+38.11	+38.17	+38.22	+38.28
99	37.22	37.28	37.33	37.39	37.44	37.50	37.56	37.61	37.67	37.72
98	36.67	36.72	36.78	36.83	37.89	36.94	37.00	37.06	37.11	37.17
97	36.11	36.17	36.22	36.28	36.33	36.39	36.44	36.50	36.56	36.61
96	35.56	35.61	35.67	35.72	35.78	35.83	35.89	35.94	36.00	36.06
+ 95	+35.00	+35.06	+35.11	+35.17	+35.22	+35.28	+35.33	+35.39	+35.44	+35.50
94	34.44	34.50	34.56	34.61	34.67	34.72	34.78	34.83	34.89	34.94
93	33.89	33.94	34.00	34.06	34.11	34.17	34.22	34.28	34.33	34.39
92	33.33	33.39	33.44	33.50	33.56	33.61	33.67	33.72	33.78	33.83
91	32.78	32.83	32.89	32.94	33.00	33.06	33.11	33.17	33.22	33.28
+ 90	+32.22	+32.28	+32.33	+32.39	+32.44	+32.50	+32.56	+32.61	+32.67	+32.72
89	31.67	31.72	31.78	31.83	31.89	31.94	32.00	32.06	32.11	32.17
88	31.11	31.17	31.22	31.28	31.33	31.39	31.44	31.50	31.56	31.61
87	30.56	30.61	30.67	30.72	30.78	30.83	30.89	30.94	31.00	31.06
86	30.00	30.06	30.11	30.17	30.22	30.28	30.33	30.39	30.44	30.50
+ 85	+29.44	+29.50	+29.56	+29.61	+29.67	+29.72	+29.78	+29.83	+29.89	+29.94
84	28.89	28.94	29.00	29.06	29.11	29.17	29.22	29.28	29.33	29.39
83	28.33	28.39	28.44	28.50	28.56	28.61	28.67	28.72	28.78	28.83
82	27.78	27.83	27.89	27.94	28.00	28.06	28.11	28.17	28.22	28.28
81	27.22	27.28	27.33	27.39	27.44	27.50	27.56	27.61	27.67	27.72
+ 80	+26.67	+26.72	+26.78	+26.83	+26.89	+26.94	+27.00	+27.06	+27.11	+27.17
79	26.11	26.17	26.22	26.28	26.33	26.39	26.44	26.50	26.56	26.61
78	25.56	25.61	25.67	25.72	25.78	25.83	25.89	25.94	26.00	26.06
77	25.00	25.06	25.11	25.17	25.22	25.28	25.33	25.39	25.44	25.50
76	24.44	24.50	24.56	24.61	24.67	24.72	24.78	24.83	24.89	24.94
+ 75	+23.89	+23.94	+24.00	+24.06	+24.11	+24.17	+24.22	+24.28	+24.33	+24.39
74	23.33	23.39	23.44	23.50	23.56	23.61	23.67	23.72	23.78	23.83
73	22.78	22.83	22.89	22.94	23.00	23.06	23.11	23.17	23.22	23.28
72	22.22	22.28	22.33	22.39	22.44	22.50	22.56	22.61	22.67	22.72
71	21.67	21.72	21.78	21.83	21.89	21.94	22.00	22.06	22.11	22.17
+ 70	+21.11	+21.17	+21.22	+21.28	+21.33	+21.39	+21.44	+21.50	+21.56	+21.61
69	20.56	20.61	20.67	20.72	20.78	20.83	20.89	20.94	21.00	21.06
68	20.00	20.06	20.11	20.17	20.22	20.28	20.33	20.39	20.44	20.50
67	19.44	19.50	19.56	19.61	19.67	19.72	19.78	19.83	19.89	19.94
66	18.89	18.94	19.00	19.06	19.11	19.17	19.22	19.28	19.33	19.39
+ 65	+18.33	+18.39	+18.44	+18.50	+18.56	+18.61	+18.67	+18.72	+18.78	+18.82
64	17.78	17.83	17.89	17.94	18.00	18.06	18.11	18.17	18.22	18.28
63	17.22	17.28	17.33	17.39	17.44	17.50	17.56	17.61	17.67	17.72
62	16.67	16.72	16.78	16.83	16.89	16.94	17.00	17.06	17.11	17.17
61	16.11	16.17	16.22	16.28	16.33	16.39	16.44	16.50	16.56	16.61

TABLE A-3
Centigrade to Fahrenheit Temperatures

°C.	.0	.1	.2	.3	.4	.5	.6	.7	.8	.9
	°F.	°F.	°F.	°F.	°F.	°F.	°F.	°F.	°F.	°F.
+60	+140.00	+140.18	+140.36	+140.54	+140.72	+140.90	+141.08	+141.26	+141.44	+141.62
59	138.20	138.38	138.56	138.74	138.92	139.10	139.28	139.46	139.64	139.82
58	136.40	136.58	136.76	136.94	137.12	137.30	137.48	137.66	137.84	138.02
57	134.60	134.78	134.96	135.14	135.32	135.50	135.68	135.86	136.04	136.22
56	132.80	132.98	133.16	133.34	133.52	133.70	133.88	134.06	134.24	134.42
+55	+131.00	+131.18	+131.36	+131.54	+131.72	+131.90	+132.08	+132.26	+132.44	+132.62
54	129.20	129.38	129.56	129.74	129.92	130.10	130.28	130.46	130.64	130.82
53	127.40	127.58	127.76	127.94	128.12	128.30	128.48	128.66	128.84	129.02
52	125.60	125.78	125.96	126.14	126.32	126.50	126.68	126.86	127.04	127.22
51	123.80	123.98	124.16	124.34	124.52	124.70	124.88	125.06	125.24	125.42
+50	+122.00	+122.18	+122.36	+122.54	+122.72	+122.90	+123.08	+123.26	+123.44	+123.62
49	120.20	120.38	120.56	120.74	120.92	121.10	121.28	121.46	121.64	121.82
48	118.40	118.58	118.76	118.94	119.12	119.30	119.48	119.66	119.84	120.02
47	116.60	116.78	116.96	117.14	117.32	117.50	117.68	117.86	118.04	118.22
46	114.80	114.98	115.16	115.34	115.52	115.70	115.88	116.06	116.24	116.42
+45	+113.00	+113.18	+113.36	+113.54	+113.72	+113.90	+114.08	+114.26	+114.44	+114.62
44	111.20	111.38	111.56	111.74	111.92	112.10	112.28	112.46	112.64	112.82
43	109.40	109.58	109.76	109.94	110.12	110.30	110.48	110.66	110.84	111.02
42	107.60	107.78	107.96	108.14	108.32	108.50	108.68	108.86	109.04	109.22
41	105.80	105.98	106.16	106.34	106.52	106.70	106.88	107.06	107.24	107.42
+40	+104.00	+104.18	+104.36	+104.54	+104.72	+104.90	+105.08	+105.26	+105.44	+105.62
39	102.20	102.38	102.56	102.74	102.92	103.10	103.28	103.46	103.64	103.82
38	100.40	100.58	100.76	100.94	101.12	101.30	101.48	101.66	101.84	102.02
37	98.60	98.78	98.96	99.14	99.32	99.50	99.68	99.86	100.04	100.22
36	96.80	96.98	97.16	97.34	97.52	97.70	97.88	98.06	98.24	98.42
+35	+ 95.00	+ 95.18	+ 95.36	+ 95.54	+ 95.72	+ 95.90	+ 96.08	+ 96.26	+ 96.44	+ 96.62
34	93.20	93.38	93.56	93.74	93.92	94.10	94.28	94.46	94.64	94.82
33	91.40	91.58	91.76	91.94	92.12	92.30	92.48	92.66	92.84	93.02
32	89.60	89.78	89.96	90.14	90.32	90.50	90.68	90.86	91.04	91.22
31	87.80	87.98	88.16	88.34	88.52	88.70	88.88	89.06	89.24	89.42
+30	+ 86.00	+ 86.18	+ 86.36	+ 86.54	+ 86.72	+ 86.90	+ 87.08	+ 87.26	+ 87.44	+ 87.62
29	84.20	84.38	84.56	84.74	84.92	85.10	85.28	85.46	85.64	85.82
28	82.40	82.58	82.76	82.94	83.12	83.30	83.48	83.66	83.84	84.02
27	80.60	80.78	80.96	81.14	81.32	81.50	81.68	81.86	82.04	82.22
26	78.80	78.98	79.16	79.34	79.52	79.70	79.88	80.06	80.24	80.42
+25	+ 77.00	+ 77.18	+ 77.36	+ 77.54	+ 77.72	+ 77.90	+ 78.08	+ 78.26	+ 78.44	+ 78.62
24	75.20	75.38	75.56	75.74	75.92	76.10	76.28	76.46	76.64	76.82
23	73.40	73.58	73.76	73.94	74.12	74.30	74.48	74.66	74.84	75.02
22	71.60	71.78	71.96	72.14	72.32	72.50	72.68	72.86	73.04	73.22
21	69.80	69.98	70.16	70.34	70.52	70.70	70.88	71.06	71.24	71.42
+20	+ 68.00	+ 68.18	+ 68.36	+ 68.54	+ 68.72	+ 68.90	+ 69.08	+ 69.26	+ 69.44	+ 69.62
19	66.20	66.38	66.56	66.74	66.92	67.10	67.28	67.46	67.64	67.82
18	64.40	64.58	64.76	64.94	65.12	65.30	65.48	65.66	65.84	66.02
17	62.60	62.78	62.96	63.14	63.32	63.50	63.68	63.86	64.04	64.22
16	60.80	60.98	61.16	61.34	61.52	61.70	61.88	62.06	62.24	62.42
+15	+ 59.00	+ 59.18	+ 59.36	+ 59.54	+ 59.72	+ 59.90	+ 60.08	+ 60.26	+ 60.44	+ 60.62
14	57.20	57.38	57.56	57.74	57.92	58.10	58.28	58.46	58.64	58.82
13	55.40	55.58	55.76	55.94	56.12	56.30	56.48	56.66	56.84	57.02
12	53.60	53.78	53.96	54.14	54.32	54.50	54.68	54.86	55.04	55.22
11	51.80	51.98	52.16	52.34	52.52	52.70	52.88	53.06	53.24	53.42

TABLE A-3 (Continued)
Centigrade to Fahrenheit Temperatures

°C.	.0	.1	.2	.3	.4	.5	.6	.7	.8	.9
	°F.	°F.	°F.	°F.	°F.	°F.	°F.	°F.	°F.	°F.
+10	+50.00	+50.18	+50.36	+50.54	+50.72	+50.90	+51.08	+51.26	+51.44	+51.62
+ 9	+48.20	+48.38	+48.56	+48.74	+48.92	+49.10	+49.28	+49.46	+49.64	+49.82
8	46.40	46.58	46.76	46.94	47.12	47.30	47.48	47.66	47.84	48.02
7	44.60	44.78	44.96	45.14	45.32	45.50	45.68	45.86	46.04	46.22
6	42.80	42.98	43.16	43.34	43.52	43.70	43.88	44.06	44.24	44.42
5	41.00	41.18	41.36	41.54	41.72	41.90	42.08	42.26	42.44	42.62
+ 4	+39.20	+39.38	+39.56	+39.74	+39.92	+40.10	+40.28	+40.46	+40.64	+40.82
3	37.40	37.58	37.76	37.94	38.12	38.30	38.48	38.66	38.84	39.02
2	35.60	35.78	35.96	36.14	36.32	36.50	36.68	36.86	37.04	37.22
1	33.80	33.98	34.16	34.34	34.52	34.70	34.88	35.06	35.24	35.42
+ 0	32.00	32.18	32.36	32.54	32.72	32.90	33.08	33.26	33.44	33.62
− 0	+32.00	+31.82	+31.64	+31.46	+31.28	+31.10	+30.92	+30.74	+30.56	+30.38
1	30.20	30.02	29.84	29.66	29.48	29.30	29.12	28.94	28.76	28.58
2	28.40	28.22	28.04	27.86	27.68	27.50	27.32	27.14	26.96	26.78
3	26.60	26.42	26.24	26.06	25.88	25.70	25.52	25.34	25.16	24.98
4	24.80	24.62	24.44	24.26	24.08	23.90	23.72	23.54	23.36	23.18
− 5	+23.00	+22.82	+22.64	+22.46	+22.28	+22.10	+21.92	+21.74	+21.56	+21.38
6	21.20	21.02	20.84	20.66	20.48	20.30	20.12	19.94	19.76	19.58
7	19.40	19.22	19.04	18.86	18.68	18.50	18.32	18.14	17.96	17.78
8	17.60	17.42	17.24	17.06	16.88	16.70	16.52	16.34	16.16	15.98
9	15.80	15.62	15.44	15.26	15.08	14.90	14.72	14.54	14.36	14.18
−10	+14.00	+13.82	+13.64	+13.46	+13.28	+13.10	+12.92	+12.74	+12.56	+12.38
11	12.20	12.02	11.84	11.66	11.48	11.30	11.12	10.94	10.76	10.58
12	10.40	10.22	10.04	9.86	9.68	9.50	9.32	9.14	8.96	8.78
13	8.60	8.42	8.24	8.06	7.88	7.70	7.52	7.34	7.16	6.98
14	6.80	6.62	6.44	6.26	6.08	5.90	5.72	5.54	5.36	5.18
−15	+ 5.00	+ 4.82	+ 4.64	+ 4.46	+ 4.28	+ 4.10	+ 3.92	+ 3.74	+ 3.56	+ 3.38
16	+ 3.20	+ 3.02	+ 2.84	+ 2.66	+ 2.48	+ 2.30	+ 2.12	+ 1.94	+ 1.76	+ 1.58
17	+ 1.40	+ 1.22	+ 1.04	+ 0.86	+ 0.68	+ 0.50	+ 0.32	+ 0.14	− 0.04	− 0.22
18	− 0.40	− 0.58	− 0.76	− 0.94	− 1.12	− 1.30	− 1.48	− 1.66	− 1.84	− 2.02
19	− 2.20	− 2.38	− 2.56	− 2.74	− 2.92	− 3.10	− 3.28	− 3.46	− 3.64	− 3.82
−20	− 4.00	− 4.18	− 4.36	− 4.54	− 4.72	− 4.90	− 5.08	− 5.26	− 5.44	− 5.62
21	5.80	5.98	6.16	6.34	6.52	6.70	6.88	7.06	7.24	7.42
22	7.60	7.78	7.96	8.14	8.32	8.50	8.68	8.86	9.04	9.22
23	9.40	9.58	9.76	9.94	10.12	10.30	10.48	10.66	10.84	11.02
24	11.20	11.38	11.56	11.74	11.92	12.10	12.28	12.46	12.64	12.82
−25	−13.00	−13.18	−13.36	−13.54	−13.72	−13.90	−14.08	−14.26	−14.44	−14.62
26	14.80	14.98	15.16	15.34	15.52	15.70	15.88	16.06	16.24	16.42
27	16.60	16.78	16.96	17.14	17.32	17.50	17.68	17.86	18.04	18.22
28	18.40	18.58	18.76	18.94	19.12	19.30	19.48	19.66	19.84	20.02
29	20.20	20.38	20.56	20.74	20.92	21.10	21.28	21.46	21.64	21.82
−30	−22.00	−22.18	−22.36	−22.54	−22.72	−22.90	−23.08	−23.26	−23.44	−23.62
31	23.80	23.98	24.16	24.34	24.52	24.70	24.88	25.06	25.24	25.42
32	25.60	25.78	25.96	26.14	26.32	26.50	26.68	26.86	27.04	27.22
33	27.40	27.58	27.76	27.94	28.12	28.30	28.48	28.66	28.84	29.02
34	29.20	29.38	29.56	29.74	29.92	30.10	30.28	30.46	30.64	30.82
−35	−31.00	−31.18	−31.36	−31.54	−31.72	−31.90	−32.08	−32.26	−32.44	−32.62
36	32.80	32.98	33.16	33.34	33.52	33.70	33.88	34.06	34.24	34.42
37	34.60	34.78	34.96	35.14	35.32	35.50	35.68	35.86	36.04	36.22
38	36.40	36.58	36.76	36.94	37.12	37.30	37.48	37.66	37.84	38.02
39	38.20	38.38	38.56	38.74	38.92	39.10	39.28	39.46	39.64	39.82

TABLE A–4

Meters into Feet

1 meter = 39.3700 inches = 3.280833 feet

Meters	0	1	2	3	4	5	6	7	8	9
	Feet	Feet	Feet	Feet	Feet	Feet	Feet	Feet	Feet	Feet
0	0.00	3.28	6.56	9.84	13.12	16.40	19.68	22.97	26.25	29.53
10	32.81	36.09	39.37	42.65	45.93	49.21	52.49	55.77	59.05	62.34
20	65.62	68.90	72.18	75.46	78.74	82.02	85.30	88.58	91.86	95.14
30	98.42	101.71	104.99	108.27	111.55	114.83	118.11	121.39	124.67	127.95
40	131.23	134.51	137.79	141.08	144.36	147.64	150.92	154.20	157.48	160.76
50	164.04	167.32	170.60	173.88	177.16	180.45	183.73	187.01	190.29	193.57
60	196.85	200.13	203.41	206.69	209.97	213.25	216.53	219.82	223.10	226.38
70	229.66	232.94	236.22	239.50	242.78	246.06	249.34	252.62	255.90	259.19
80	262.37	265.75	269.03	272.31	275.59	278.87	282.15	285.43	288.71	291.99
90	295.27	298.56	301.84	305.12	308.40	311.68	314.96	318.24	321.52	324.80
100	328.08	331.36	334.64	337.93	341.21	344.49	347.77	351.05	354.33	357.61
110	360.89	364.17	367.45	370.73	374.01	377.30	380.58	383.86	387.14	390.42
120	393.70	396.98	400.26	403.54	406.82	410.10	413.38	416.67	419.95	423.23
130	426.51	429.79	433.07	436.35	439.63	442.91	446.19	449.47	452.75	456.04
140	459.32	462.60	465.88	469.16	472.44	475.72	479.00	482.28	485.56	488.84
150	492.12	495.41	498.69	501.97	505.25	508.53	511.81	515.09	518.37	521.65
160	524.93	528.21	531.49	534.78	538.06	541.34	544.62	547.90	551.18	554.46
170	557.74	561.02	564.30	567.58	570.86	574.15	577.43	580.71	583.99	587.27
180	590.55	593.83	597.11	600.39	603.67	606.95	610.23	613.52	616.80	620.08
190	623.36	626.64	629.92	633.20	636.48	639.76	643.04	646.32	649.60	652.89
200	656.17	659.45	662.73	666.01	669.29	672.57	675.85	679.13	682.41	685.69
210	688.97	692.26	695.54	698.82	702.10	705.38	708.66	711.94	715.22	718.50
220	721.78	725.06	728.34	731.63	734.91	738.19	741.47	744.75	748.03	751.31
230	754.59	757.87	761.15	764.43	767.71	771.00	774.28	777.56	780.84	784.12
240	787.40	790.68	793.96	797.24	800.52	803.80	807.09	810.37	813.65	816.93
250	820.21	823.49	826.77	830.05	833.33	836.61	839.89	843.17	846.45	849.74
260	853.02	856.30	859.58	862.86	866.14	869.42	872.70	875.98	879.26	882.54
270	885.82	889.11	892.39	895.67	898.95	902.23	905.51	908.79	912.07	915.35
280	918.63	921.91	925.19	928.48	931.76	935.04	938.32	941.60	944.88	948.16
290	951.44	954.72	958.00	961.28	964.56	967.85	971.13	974.41	977.69	980.97
300	984.25	987.53	990.81	994.09	997.37	1000.65	1003.93	1007.22	1010.50	1013.78
310	1017.06	1020.34	1023.62	1026.90	1030.18	1033.46	1036.74	1040.02	1043.30	1046.59
320	1049.87	1053.15	1056.43	1059.71	1062.99	1066.27	1069.55	1072.83	1076.11	1079.39
330	1082.67	1085.96	1089.24	1092.52	1095.80	1099.08	1102.36	1105.64	1109.92	1112.20
340	1115.48	1118.76	1122.04	1125.33	1128.61	1131.89	1135.17	1138.45	1141.73	1145.01
350	1148.29	1151.57	1154.85	1158.13	1161.41	1164.70	1167.98	1171.26	1174.54	1177.82
360	1181.10	1184.38	1187.66	1190.94	1194.22	1197.50	1200.78	1204.07	1207.35	1210.63
370	1213.91	1217.19	1220.47	1223.75	1227.03	1230.31	1233.59	1236.87	1240.15	1243.44
380	1246.72	1250.00	1253.28	1256.56	1259.84	1263.12	1266.40	1269.68	1272.96	1276.24
390	1279.52	1282.81	1286.09	1289.37	1292.65	1295.93	1299.21	1302.49	1305.77	1309.05
400	1312.33	1315.61	1318.89	1322.18	1325.46	1328.74	1332.02	1335.30	1338.58	1341.86
410	1345.14	1348.42	1351.70	1354.98	1358.26	1361.55	1364.83	1368.11	1371.39	1374.67
420	1377.95	1381.23	1384.51	1387.79	1391.07	1394.35	1397.63	1400.92	1404.20	1407.48
430	1410.76	1414.04	1417.32	1420.60	1423.88	1427.16	1430.44	1433.72	1437.00	1440.29
440	1443.57	1446.85	1450.13	1453.41	1456.69	1459.97	1463.25	1466.53	1469.81	1473.09
450	1476.37	1479.66	1482.94	1486.22	1489.50	1492.78	1496.06	1499.34	1502.62	1505.90
460	1509.18	1512.46	1515.74	1519.03	1522.31	1525.59	1528.87	1532.15	1535.43	1538.71
470	1541.99	1545.27	1548.55	1551.83	1555.11	1558.40	1561.68	1564.96	1568.24	1571.52
480	1574.80	1578.08	1581.36	1584.64	1587.92	1591.20	1594.48	1597.77	1601.05	1604.33
490	1607.61	1610.89	1614.17	1617.45	1620.73	1624.01	1627.29	1630.57	1633.85	1637.14

RELATIONSHIP OF RELATIVE HUMIDITY AND DEW POINT
WITH RESPECT TO ICE AND WATER RESPECTIVELY

Effective January 1, 1949, the International Meteorological Organization
adopted dew point and relative humidity with respect to water at temperatures below 32° F. The following tables show the relationship to values
used before, for humidity and dew point with respect to ice.

R.H. denotes relative humidity; D.P. denotes dew point; subscript w denotes
"with respect to water"; subscript i denotes "with respect to ice."

TABLE A–5A

When Saturation Exists with Respect to Water

Temperature ($°F.$)	$R.H._w$ ($\%$)	$R.H._i$ ($\%$)	$D.P._w$ ($°F.$)	$D.P._i$ ($°F.$)
32	100	100	32	32
20	100	106.7	20	21.4
0	100	118.9	0	+3.3
−20	100	132.5	−20	−15.1
−40	100	147.4	−40	−33.8
−60	100	163.1	−60	−52.8

TABLE A–5B

When Saturation Exists with Respect to Ice

Temperature ($°F.$)	$R.H._i$ ($\%$)	$R.H._w$ ($\%$)	$D.P._i$ ($°F.$)	$D.P._w$ ($°F.$)
32	100	100	32	32
20	100	93.7	20	18.5
0	100	84.1	0	−3.7
−20	100	75.5	−20	−25.4
−40	100	67.8	−40	−46.6
−60	100	61.3	−60	

The values of dew point and relative humidity obtained from the standard
Psychometric Tables and Tables A–7 and A–8 must be corrected when tem-
perature or dew point is below 32° F. to values with respect to water. Tables
A–5C and A–5D may be used for this purpose.

TABLE A–5C

Relative Humidity Conversion Table

(Tabular values are relative humidities with respect to water (R.H.$_w$) corresponding to relative humidities with respect to ice (R.H.$_i$) given at heads of the columns.)

Dry Bulb Temp. (°F.)	Relative Humidity with Respect to Ice (R.H.$_i$)										Dry Bulb Temp. (°F.)	Relative Humidity Ratio R.H.$_w$/R.H.$_i$
	10%	20%	30%	40%	50%	60%	70%	80%	90%	100%		
	%	%	%	%	%	%	%	%	%	%		
32	10.0	20.0	30.0	40.0	50.0	60.0	70.0	80.0	90.0	100.0	32	0.9999
30	9.9	19.8	29.7	39.6	49.5	59.4	69.2	79.1	89.0	98.9	30	.9892
20	9.4	18.7	28.1	37.5	46.9	56.2	65.6	75.0	84.3	93.7	20	.9371
10	8.9	17.8	26.6	35.5	44.4	53.3	62.1	71.0	79.9	88.8	10	.8876
0	8.4	16.8	25.2	33.6	42.1	50.5	58.9	67.3	75.7	84.1	0	.8410
−10	8.0	15.9	23.9	31.9	39.8	47.8	55.7	63.7	71.7	79.6	−10	.7963
−20	7.5	15.1	22.6	30.2	37.7	45.3	52.8	60.4	67.9	75.5	−20	.7546
−30	7.2	14.3	21.5	28.6	35.8	42.9	50.1	57.2	64.4	71.5	−30	.7152
−40	6.8	13.6	20.4	27.1	33.9	40.7	47.5	54.3	61.1	67.8	−40	.6785
−50	6.4	12.9	19.3	25.8	32.2	38.7	45.1	51.5	58.0	64.4	−50	.6444
−60	6.1	12.3	18.4	24.5	30.7	36.8	42.9	49.0	55.2	61.3	−60	.6130

Note: Saturation vapor pressures over ice and water, used in computing this table, are based on formulas by J. A. Goff and S. Gratch, *Trans. Amer. Soc. Heat. and Vent. Eng.,* Vol. 52, page 95 (1946). Formula for saturation vapor pressure over water assumed to apply from −60° F. to 140° F.

TABLE A–5D

DEW POINT CONVERSION SCALE
SHOWING RELATIONSHIP BETWEEN
DEW POINT WITH RESPECT TO ICE AND DEW POINT WITH RESPECT TO WATER
(FAHRENHEIT TEMPERATURES)

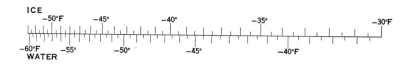

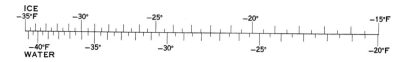

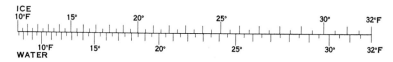

Note: Saturation vapor pressures over ice and water, used in computing this scale, are based on formulas by J. A. Goff and S. Gratch, *Trans. Amer. Soc. Heat. and Vent. Eng.*, Vol. 52, page 95 (1946). Formula for saturation vapor pressure over water assumed to apply from −60° F. to 140° F. Courtesy U. S. Weather Bureau.

PSYCHROMETRIC TABLES

Abridged from *U. S. Weather Bureau Publication* 235 by C. F. Marvin

Corrections for pressures other than 29 inches

1. Find dew point and relative humidity using table for 29 inches (1,000 feet altitude).
2. Add corrections at pressures less than 29 inches (higher elevations through 1,000 feet) and subtract corrections at pressures higher than 29 inches (sea level to 1,000 feet elevation).

TABLE A–6

Air Temperature of	Depression of Wet Bulb	30 in. (sea level)				27 in. (3,000 ft.)				25 in. (5,000 ft.)			
		1	2	5	10	1	2	5	10	1	2	5	10
20	Dew-point correction, °F.	−1	−1	−2		0	0	+2		0	0	+4	
	Relative-humidity correction, %	0	0	−2		+1	+2	+3		+2	+3	+7	
40	Dew-point correction, °F.	0	0	0	−1	0	+1	+1	+3	0	+1	+1	+5
	Relative-humidity correction, %	0	−1	−1	−1	0	0	+1	+3	0	+1	+3	+6
60	Dew-point correction, °F.	0	0	0	−1	0	0	+1	0	+1	0	+1	+2
	Relative-humidity correction, %	0	0	0	−1	+1	0	+1	+1	+1	+1	+1	+3
80	Dew-point correction, °F.	0	0	0	0	0	0	0	+1	0	0	0	+1
	Relative-humidity correction, %	0	0	0	0	0	+1	+1	+1	0	+1	+1	+1

Table A–7 for Temperature of the Dew Point and Table A–8 for Relative Humidity are computed with respect to ice below 32° F. Tabular Values of Relative Humidity, Vapor Pressure, and Dew Point are with respect to ice below 32° F. and with respect to water above 32° F. Depression of the Wet-Bulb is for an ice-covered bulb below 32° F.

Appendix

TABLE A–7

Temperature of Dew Point in Degrees Fahrenheit

[Pressure = 29.0 inches]

Air Temperature t	Vapor Pressure e	— Depression of Wet-Bulb Thermometer (t − t′)													
		0.2	0.4	0.6	0.8	1.0	1.2	1.4	1.6	1.8	2.0	2.2	2.4	2.6	2.8	3.0
−40	0.0039	−51														
−39	.0041	−50														
−38	.0044	−49														
−37	.0046	−47														
−36	.0048	−46														
−35	.0051	−44														
−34	.0054	−43	−58													
−33	.0057	−41	−55													
−32	.0061	−40	−52													
−31	.0065	−38	−49													
−30	.0069	−36	−47													
−29	.0074	−34	−44													
−28	.0078	−33	−42	−56												
−27	.0083	−32	−40	−52												
−26	.0089	−30	−37	−49												
−25	.0094	−29	−35	−45												
−24	.0100	−28	−34	−42	−57											
−23	.0106	−27	−32	−40	−51											
−22	.0112	−26	−30	−37	−47											
−21	.0119	−24	−29	−34	−44	−60										
−20	.0126	−23	−28	−33	−40	−53										
−19	.0133	−22	−26	−31	−37	−48										
−18	.0141	−21	−25	−29	−34	−44										
−17	.0150	−20	−23	−27	−32	−40	−55									
−16	.0159	−18	−22	−26	−30	−37	−48									
−15	.0168	−17	−20	−24	−28	−34	−43	−59								
−14	.0178	−16	−19	−22	−26	−31	−39	−51								
−13	.0188	−15	−18	−21	−24	−29	−35	−45								
−12	.0199	−14	−17	−19	−23	−27	−32	−40	−53							
−11	.0210	−13	−15	−18	−21	−25	−29	−36	−46							
−10	.0222	−12	−14	−17	−19	−23	−27	−32	−40	−56						
−9	.0234	−11	−13	−15	−18	−21	−25	−29	−36	−47						
−8	.0247	−10	−12	−14	−16	−19	−23	−27	−32	−41	−55					
−7	.0260	−9	−11	−13	−15	−18	−21	−24	−29	−35	−46					
−6	.0275	−8	−10	−12	−14	−16	−19	−22	−26	−32	−40	−54				
−5	.0291	−7	−8	−10	−12	−15	−17	−20	−24	−29	−35	−45				
−4	.0307	−5	−7	−9	−11	−13	−16	−18	−22	−26	−30	−38	−50			
−3	.0325	−4	−6	−8	−10	−12	−14	−16	−19	−23	−27	−32	−42	−59		
−2	.0344	−3	−5	−6	−8	−10	−12	−14	−17	−20	−24	−29	−35	−47		
−1	.0363	−2	−4	−5	−7	−9	−11	−12	−15	−18	−21	−25	−30	−38	−51	
0	.0383	−1	−3	−4	−5	−7	−9	−11	−13	−16	−19	−22	−26	−32	−40	−57
1	.0403	±0	−2	−3	−4	−6	−7	−9	−12	−14	−16	−19	−23	−28	−34	−44
2	.0423	+1	±0	−2	−3	−5	−6	−8	−10	−12	−14	−17	−20	−24	−29	−35
3	.0444	2	+1	−1	−2	−3	−5	−6	−8	−10	−12	−15	−18	−21	−25	−30
4	.0467	3	2	±0	−1	−2	−4	−5	−7	−9	−11	−13	−15	−18	−22	−26
5	.0491	4	3	+2	±0	−1	−2	−4	−5	−7	−9	−11	−13	−16	−19	−22
6	.0515	5	4	3	+1	±0	−1	−2	−4	−5	−7	−9	−11	−14	−16	−19
7	.0542	6	5	4	3	+1	±0	−1	−3	−4	−5	−7	−9	−12	−14	−16
8	.0570	7	6	5	4	3	+1	±0	−1	−3	−4	−5	−7	−10	−12	−14
9	.0600	8	7	6	5	4	3	+1	±0	−1	−3	−4	−5	−7	−10	−12
10	.0631	9	8	7	6	5	4	3	+1	±0	−1	−2	−4	−6	−8	−10
11	.0665	10	9	8	7	6	5	4	3	+2	±0	−1	−2	−4	−6	−7
12	.0699	11	10	9	8	7	6	5	4	3	+2	+1	−1	−2	−4	−5
13	.0735	12	11	11	10	9	8	7	6	5	3	+2	+1	−1	−2	−4
14	.0772	13	12	12	11	10	9	8	7	6	5	4	3	+1	±0	−2
15	.0810	14	13	13	12	11	10	9	8	7	6	5	4	3	+1	±0
16	.0850	15	14	14	13	12	11	10	9	9	8	7	6	5	4	+2
17	.0891	16	16	15	14	13	12	11	11	10	9	8	7	6	5	4
18	.0933	17	17	16	15	14	13	12	12	11	10	9	8	7	6	5
19	.0979	18	18	17	16	15	14	13	13	12	11	11	10	9	8	7
20	.1026	19	19	18	17	17	16	15	14	13	13	12	11	10	9	8

Embedded auxiliary table (low temperatures, small depression (t − t′) = .1 .2 .3 .4 .5):

t	e	t	e
−60	0.0010	−50	0.0021
−59	.0011	−49	.0022
−58	.0012	−48	.0024
−57	.0013	−47	.0026
−56	.0013	−46	.0027
−55	.0015	−45	.0029
−54	.0016	−44	.0031
−53	.0017	−43	.0033
−52	.0018	−42	.0035
−51	.0019	−41	.0037
		−40	.0039
		−39	.0041
		−38	.0044
		−37	.0046
		−36	.0048
		−35	.0051
		−34	.0054
		−33	.0057
		−32	.0061
		−31	.0065
		−30	.0069

Dew point at depression (t − t′):

t	.1	.2	.3	.4	.5
−60	−60				
−59	−58				
−58	−56				
−57	−55				
−56	−53				
−55	−51	−59			
−54	−50	−57			
−53	−49	−55			
−52	−48	−55			
−51	−46	−53			
−45	−51	−60			
−44	−50	−58			
−43	−49	−58			
−42	−47	−55			
−41	−46	−53			
−40	−45	−51			
−39	−44	−50			
−38	−43	−49			
−37	−41	−47			
−36	−40	−45			
−35	−44	−51	−58		
−34	−38	−43	−49	−58	
−33	−37	−41	−47	−55	
−32	−35	−40	−45	−52	
−31	−34	−33	−43	−49	−58
−30	−36	−41	−47	−54	

Note: Vapor pressure e in inches of mercury.

TABLE A–7 (Continued)

Temperature of Dew Point in Degrees Fahrenheit

[Pressure = 29.0 inches]

Air Temperature t	Vapor Pressure e	Depression of Wet-Bulb Thermometer ($t - t'$)															
		0.5	1.0	1.5	2.0	2.5	3.0	3.5	4.0	4.5	5.0	5.5	6.0	6.5	7.0	7.5	8.0
20	0.103	18	17	15	13	11	8	5	2	−1	−5	−11	−18	−30			
21	.108	19	18	16	14	12	10	7	4	+1	−3	−8	−14	−23	−42		
22	.113	20	19	17	15	13	11	9	6	3	−1	−5	−10	−17	−29		
23	.118	22	20	18	16	14	12	10	8	5	+1	−3	−7	−13	−22	−40	
24	.124	23	21	19	18	16	14	11	9	6	3	±0	−4	−10	−17	−28	
25	0.130	24	22	21	19	17	15	13	11	8	5	+2	−2	−6	−12	−21	−36
26	.136	25	23	22	20	18	16	14	12	10	7	4	±0	−4	−9	−15	−26
27	.143	26	24	23	21	20	18	16	14	12	9	6	+3	−1	−5	−11	−19
28	.150	27	25	24	22	21	19	17	15	13	11	8	5	+2	−2	−7	−14
29	.157	28	26	25	24	22	20	19	17	15	12	10	7	4	±0	−4	−9
30	0.164	29	27	26	25	23	22	20	18	16	14	12	9	6	+3	−1	−5
31	.172	30	29	27	26	24	23	21	20	18	16	13	11	8	5	+2	−2
32	.180	31	30	28	27	26	24	23	21	19	17	15	13	10	8	4	+1
33	.187	32	31	29	28	27	25	24	22	21	19	17	15	12	10	7	3
34	.195	33	32	30	29	28	27	25	24	22	20	18	16	14	12	9	6
35	0.203	34	33	31	30	29	28	26	25	23	22	20	18	16	14	11	8
36	.211	35	34	32	31	30	29	27	26	25	23	21	20	18	15	13	11
37	.219	36	35	33	32	31	30	28	27	26	24	23	21	19	17	15	13
38	.228	37	36	34	33	32	31	30	28	27	26	24	23	21	19	17	14
39	.237	38	37	36	34	33	32	31	29	28	27	25	24	22	21	19	16
40	0.247	39	38	37	35	34	33	32	31	29	28	27	25	23	22	20	18
41	.256	40	39	38	37	35	34	33	32	30	29	28	26	25	23	22	20
42	.266	41	40	39	38	36	35	34	33	32	30	29	28	26	25	23	21
43	.277	42	41	40	39	38	36	35	34	33	31	30	29	27	26	24	23
44	.287	43	42	41	40	39	38	36	35	34	32	31	30	29	27	26	24
45	0.298	44	43	42	41	40	39	37	36	35	34	32	31	30	29	27	26
46	.310	45	44	43	42	41	40	39	37	36	35	34	32	31	30	28	27
47	.322	46	45	44	43	42	41	40	39	37	36	35	34	32	31	30	28
48	.334	47	46	45	44	43	42	41	40	39	37	36	35	34	32	31	30
49	.347	48	47	46	45	44	43	42	41	40	39	37	36	35	34	32	31
50	0.360	49	48	47	46	45	44	43	42	41	40	39	37	36	35	34	32
51	.373	50	49	48	47	46	45	44	43	42	41	40	39	37	36	35	34
52	.387	51	50	49	48	47	46	45	44	43	42	41	40	39	37	36	35
53	.402	52	51	50	49	48	47	46	45	44	43	42	41	40	39	38	36
54	.417	53	52	51	50	49	48	47	46	45	44	43	42	41	40	39	38
55	0.432	54	53	52	52	51	50	49	48	47	46	45	43	42	41	40	39
56	.448	55	54	53	53	52	51	50	49	48	47	46	45	44	43	41	40
57	.465	56	55	54	54	53	52	51	50	49	48	47	46	45	44	43	42
58	.482	57	56	56	55	54	53	52	51	50	49	48	47	46	45	44	43
59	.499	58	57	57	56	55	54	53	52	51	50	49	48	47	46	45	44
60	0.517	59	58	58	57	56	55	54	53	52	51	50	49	48	47	46	45
61	.536	60	59	59	58	57	56	55	54	53	52	51	50	49	48	47	46
62	.555	61	60	60	59	58	57	56	55	55	54	53	52	51	50	49	48
63	.575	62	61	61	60	59	58	57	56	56	55	54	53	52	51	50	49
64	.595	63	62	62	61	60	59	58	58	57	56	55	54	53	52	51	50
65	0.616	64	63	63	62	61	60	59	59	58	57	56	55	54	53	52	51
66	.638	65	64	64	63	62	61	61	60	59	58	57	56	55	54	54	53
67	.661	66	65	65	64	63	62	62	61	60	59	58	57	56	56	55	54
68	.684	67	67	66	65	64	63	63	62	61	60	59	58	58	57	56	55
69	.707	68	68	67	66	65	64	64	63	62	61	60	60	59	58	57	56
70	0.732	69	69	68	67	66	66	65	64	63	62	62	61	60	59	58	57
71	.757	70	70	69	68	67	67	66	65	64	63	63	62	61	60	59	58
72	.783	71	71	70	69	68	68	67	66	65	65	64	63	62	61	60	60
73	.810	72	72	71	70	69	69	68	67	66	66	65	64	63	62	62	61
74	.838	73	73	72	71	70	70	69	68	67	67	66	65	64	64	63	62
75	0.866	74	74	73	72	71	71	70	69	68	68	67	66	65	65	64	63
76	.896	75	75	74	73	73	72	71	70	70	69	68	67	66	66	65	64
77	.926	76	76	75	74	74	73	72	71	71	70	69	68	68	67	66	65
78	.957	77	77	76	75	75	74	73	72	72	71	70	69	69	68	67	66
79	.989	78	78	77	76	76	75	74	73	73	72	71	70	70	69	68	67
80	1.022	79	79	78	77	77	76	75	75	74	73	72	72	71	70	69	69

TABLE A–7 (*Continued*)

Temperature of Dew Point in Degrees Fahrenheit

[Pressure = 29.0 inches]

Air Temperature t	Vapor Pressure e	Depression of Wet-Bulb Thermometer (t − t')															
		8.5	9.0	9.5	10.0	10.5	11.0	11.5	12.0	12.5	13.0	13.5	14.0	14.5	15.0	15.5	16.0
26	0.136	−53															
27	.143	−32															
28	.150	−23	−45														
29	.157	−17	−29														
30	0.164	−11	−20	−36													
31	.172	−7	−14	−24	−50												
32	.180	−4	−9	−16	−29												
33	.187	−1	−5	−11	−20	−36											
34	.195	+2	−2	−7	−14	−24	−50										
35	0.203	5	+1	−3	−8	−16	−28										
36	.211	8	4	±0	−4	−10	−19	−34									
37	.219	10	7	+3	−1	−6	−12	−22	−44								
38	.228	12	9	6	+3	−2	−7	−14	−25								
39	.237	14	12	9	6	+2	−3	−8	−16	−30							
40	0.247	16	14	11	8	5	+1	−4	−10	−18	−35						
41	.256	18	16	13	11	8	4	±0	−5	−11	−21	−45					
42	.266	19	17	15	13	10	7	+4	−1	−6	−13	−24	−59				
43	.277	21	19	17	15	13	10	7	+3	−2	−7	−15	−28				
44	.287	23	21	19	17	15	12	9	6	+2	−2	−8	−17	−32			
45	0.298	24	22	20	19	17	14	11	8	6	+2	−3	−9	−18	−37		
46	.310	25	24	22	20	18	16	14	11	8	5	+1	−4	−11	−20	−44	
47	.322	27	25	24	22	20	18	16	13	11	8	4	±0	−5	−12	−23	−53
48	.334	28	27	25	23	22	20	18	15	13	10	7	+4	−1	−6	−13	−25
49	.347	29	28	27	25	23	21	20	17	15	13	10	7	+3	−2	−7	−15
50	0.360	31	29	28	27	25	23	21	19	17	15	12	9	6	+2	−2	−8
51	.373	32	31	29	28	26	25	23	21	19	17	15	12	9	6	+2	−3
52	.387	34	32	31	29	28	26	25	23	21	19	17	14	12	9	5	+1
53	.402	35	34	32	31	29	28	26	24	23	21	19	16	14	11	9	5
54	.417	36	35	34	32	31	29	28	26	24	23	21	19	16	14	11	8
55	0.432	38	36	35	34	32	31	29	28	26	24	23	21	19	16	14	11
56	.448	39	38	36	35	34	32	31	29	28	26	24	23	21	19	16	14
57	.465	40	39	38	36	35	34	32	31	29	28	26	24	22	21	18	16
58	.482	42	40	39	38	36	35	34	32	31	29	28	26	24	22	20	18
59	.499	43	42	41	40	38	37	35	34	32	31	29	28	26	24	22	20
60	0.517	44	43	42	41	39	38	37	35	34	32	31	29	28	26	24	22
61	.536	45	44	43	42	41	39	38	37	35	34	32	31	29	28	26	24
62	.555	47	46	44	43	42	41	40	38	37	35	34	32	31	30	28	26
63	.575	48	47	46	45	43	42	41	40	38	37	36	34	33	31	30	28
64	.595	49	48	47	46	45	44	42	41	40	38	37	36	34	33	31	30
65	0.616	50	49	48	47	46	45	44	43	41	40	39	37	36	34	33	31
66	.638	52	51	50	49	48	47	46	45	44	43	42	40	39	37	36	33
67	.661	53	52	51	50	49	48	46	45	44	43	42	40	39	37	36	35
68	.684	54	53	52	51	50	49	48	47	46	44	43	42	41	39	38	36
69	.707	55	54	53	52	51	50	49	48	47	46	45	43	42	41	40	38
70	0.732	56	55	54	53	53	52	51	50	49	48	47	46	45	44	42	41
71	.757	58	57	56	55	55	54	53	52	51	50	49	48	47	46	44	43
72	.783	59	58	57	56	56	55	54	53	52	51	50	49	48	46	45	44
73	.810	60	59	58	58	57	56	55	54	53	52	51	50	49	48	47	46
74	.838	61	60	60	59	58	57	56	55	54	54	53	52	50	49	48	47
75	0.866	62	61	60	60	59	58	57	56	55	54	53	52	51	50	48	47
76	.896	63	62	62	61	60	59	58	57	56	55	54	53	52	51	50	49
77	.926	64	64	63	62	61	60	59	58	57	56	55	54	53	52	51	50
78	.957	66	65	64	63	62	61	60	59	58	57	56	55	54	53	52	52
79	0.989	67	66	65	64	63	62	62	61	60	59	58	57	56	55	54	53
80	1.022	68	67	66	65	64	64	63	62	61	60	59	58	57	56	55	54

TABLE A–7 (Continued)

Temperature of Dew Point in Degrees Fahrenheit

[Pressure = 29.0 inches]

Air Temperature t	Depression of Wet-Bulb Thermometer (t − t′)															
	16.5	17.0	17.5	18.0	18.5	19.0	19.5	20.0	20.5	21.0	21.5	22.0	22.5	23.0	23.5	24.0
49	−28															
50	−17	−32														
51	−10	−18	−37													
52	−4	−11	−20	−45												
53	±0	−5	−12	−23	−55											
54	+5	±0	−6	−13	−25											
55	8	+4	−1	−6	−14	−27										
56	11	7	+4	−1	−7	−16	−30									
57	13	10	7	+3	−2	−8	−17	−33								
58	16	13	10	7	+3	−2	−9	−18	−37							
59	18	16	13	10	7	+2	−3	−9	−19	−41						
60	20	18	16	13	10	6	+2	−3	−10	−20	−46					
61	22	20	18	16	13	10	6	+2	−3	−10	−21	−51				
62	24	22	20	18	16	13	10	6	+2	−4	−11	−22	−59			
63	26	24	22	20	18	16	13	10	6	+2	−4	−11	−23			
64	28	26	25	23	21	18	16	13	10	6	+1	−4	−12	−24		
65	30	28	26	25	23	21	18	16	13	10	6	+1	−4	−12	−25	
66	31	30	28	27	25	23	21	18	16	13	10	6	+1	−4	−12	−25
67	33	32	30	29	27	25	23	21	19	16	13	10	6	+1	−4	−12
68	35	33	32	30	29	27	25	23	21	19	16	13	10	6	+1	−4
69	37	35	34	32	31	29	27	25	23	21	19	16	13	10	6	+1
70	38	37	35	34	32	31	29	27	25	23	21	19	16	13	10	6
71	40	38	37	36	34	32	31	29	27	26	24	21	19	17	14	10
72	42	40	39	37	36	34	32	31	29	28	26	24	21	19	17	14
73	43	42	40	39	38	36	34	33	31	30	28	26	24	22	19	17
74	45	43	42	41	39	38	36	35	33	32	30	28	26	24	22	20
75	46	45	44	42	41	40	38	37	35	33	32	30	28	26	24	22
76	48	46	45	44	43	41	40	38	37	35	34	32	30	29	27	25
77	49	48	47	46	44	43	42	40	39	37	36	34	32	31	29	27
78	50	49	48	47	46	44	43	42	41	39	38	36	34	33	31	29
79	52	51	50	49	47	46	45	44	42	41	39	38	36	35	33	31
80	53	52	51	50	49	48	46	45	44	43	41	40	38	37	35	33

t	Depression of Wet-Bulb Thermometer (t − t′)														
	24.5	25.0	25.5	26.0	26.5	27.0	27.5	28.0	28.5	29.0	29.5	30.0	30.5	31.0	31.5
67	−26														
68	−12	−26													
69	−4	−13	−27												
70	+1	−4	−12	−27											
71	6	+2	−4	−12	−26										
72	10	7	+2	−4	−12	−26									
73	14	11	7	+2	−4	−12	−26								
74	17	14	11	7	+2	−3	−12	−25							
75	20	17	14	11	7	+3	−3	−11	−25						
76	23	20	18	15	11	8	+3	−3	−10	−24					
77	25	23	20	18	15	12	8	+3	−2	−10	−22				
78	27	25	23	21	18	15	12	8	+4	−2	−9	−21			
79	29	28	26	24	21	19	16	13	9	+4	−1	−9	−20		
80	31	30	28	26	24	22	19	16	13	9	+5	−1	−8	−19	−54

TABLE A–7 *(Continued)*

Temperature of Dew Point in Degrees Fahrenheit

[Pressure = 29.0 inches]

Air Temperature t	Vapor Pressure e	Depression of Wet-Bulb Thermometer $(t - t')$														
		1	2	3	4	5	6	7	8	9	10	11	12	13	14	15
80	1.022	79	77	76	75	73	72	70	69	67	65	64	62	60	58	56
81	.056	80	78	77	76	74	73	71	70	68	66	65	63	61	60	58
82	.091	81	79	78	77	75	74	72	71	69	68	66	64	63	61	59
83	.127	82	80	79	78	76	75	73	72	70	69	67	65	64	62	60
84	.163	83	81	80	79	77	76	74	73	71	70	68	67	65	63	61
85	1.201	84	82	81	80	78	77	75	74	72	71	69	68	66	64	63
86	.241	85	83	82	81	79	78	77	75	74	72	71	69	67	66	64
87	.281	86	84	83	82	80	79	78	76	75	73	72	70	69	67	65
88	.322	87	85	84	83	81	80	79	77	76	74	73	71	70	68	66
89	.364	88	86	85	84	82	81	80	78	77	75	74	72	71	69	68
90	1.408	89	87	86	85	84	82	81	79	78	77	75	74	72	70	69
91	.453	90	88	87	86	85	83	82	80	79	78	76	75	73	72	70
92	.499	91	90	88	87	86	84	83	82	80	79	77	76	74	73	71
93	.546	92	91	89	88	87	85	84	83	81	80	78	77	75	74	72
94	.595	93	92	90	89	88	86	85	84	82	81	80	78	77	75	74
95	1.645	94	93	91	90	89	87	86	85	83	82	81	79	78	76	75
96	.696	95	94	92	91	90	88	87	86	84	83	82	80	79	77	76
97	.749	96	95	93	92	91	90	88	87	86	84	83	81	80	78	77
98	.803	97	96	94	93	92	91	89	88	87	85	84	82	81	80	78
99	.859	98	97	95	94	93	92	90	89	88	86	85	84	82	81	79
100	1.916	99	98	96	95	94	93	91	90	89	87	86	85	83	82	80
101	.975	100	99	97	96	95	94	92	91	90	88	87	86	84	83	82
102	2.035	101	100	98	97	96	95	93	92	91	90	88	87	86	84	83
103	.097	102	101	99	98	97	96	94	93	92	91	89	88	87	85	84
104	.160	103	102	100	99	98	97	96	94	93	92	90	89	88	86	85
105	2.225	104	103	101	100	99	98	97	95	94	93	91	90	89	87	86
106	.292	105	104	102	101	100	99	98	96	95	94	92	91	90	89	87
107	.360	106	105	103	102	101	100	99	97	96	95	94	92	91	90	88
108	.431	107	106	104	103	102	101	100	98	97	96	95	93	92	91	89
109	.503	108	107	105	104	103	102	101	99	98	97	96	94	93	92	90
110	2.576	109	108	107	105	104	103	102	101	99	98	97	95	94	93	92
111	.652	110	109	108	106	105	104	103	102	100	99	99	97	95	94	93
112	.730	111	110	109	107	106	105	104	103	101	100	100	99	98	96	94
113	.810	112	111	110	108	107	106	105	104	102	101	101	100	99	97	95
114	.891	113	112	111	109	108	107	106	105	103	102	102	101	100	98	96
115	2.975	114	113	112	110	109	108	107	106	104	103	102	101	100	98	97
116	3.061	115	114	113	111	110	109	108	107	106	104	103	102	101	100	98
117	.148	116	115	114	112	111	110	109	108	107	105	104	103	102	100	99
118	.239	117	116	115	113	112	111	110	109	108	106	105	104	103	101	100
119	.331	118	117	116	114	113	112	111	110	109	107	106	105	104	103	101
120	3.425	119	118	117	115	114	113	112	111	110	108	107	106	105	104	102
121	.522	120	119	118	116	115	114	113	112	111	110	108	107	106	105	103
122	.621	121	120	119	118	116	115	114	113	112	111	109	108	107	106	105
123	.723	122	121	120	119	117	116	115	114	113	112	110	109	108	107	106
124	.827	123	122	121	120	118	117	116	115	114	113	111	110	109	108	107
125	3.933	124	123	122	121	119	118	117	116	115	114	112	111	110	109	108
126	4.042	125	124	123	122	120	119	118	117	116	115	114	112	111	110	109
127	.154	126	125	124	123	121	120	119	118	117	116	115	113	112	111	110
128	.268	127	126	125	124	122	121	120	119	118	117	116	114	113	112	111
129	.385	128	127	126	125	123	122	121	120	119	118	117	115	114	113	112
130	4.504	129	128	127	126	124	123	122	121	120	119	118	117	115	114	113
131	.627	130	129	128	127	125	124	123	122	121	120	119	118	116	115	114
132	.752	131	130	129	128	127	125	124	123	122	121	120	119	117	116	115
133	4.880	132	131	130	129	128	126	125	124	123	122	121	120	118	117	116
134	5.011	133	132	131	130	129	127	126	125	124	123	122	121	120	119	117
135	5.145	134	133	132	131	130	128	127	126	125	124	123	122	121	119	118
136	.282	135	134	133	132	131	129	128	127	126	125	124	123	122	120	119
137	.422	136	135	134	133	132	130	129	128	127	126	125	124	123	121	120
138	.565	137	136	135	134	133	131	130	129	128	127	126	125	124	122	121
139	.712	138	137	136	135	134	132	131	130	129	128	127	126	125	123	122
140	5.862	139	138	137	136	135	133	132	131	130	129	128	127	126	124	123

TABLE A–7 *(Continued)*

Temperature of Dew Point in Degrees Fahrenheit

[Pressure = 29.0 inches]

Air Temperature t	Vapor Pressure e	Depression of Wet-Bulb Thermometer (t − t')													
		16	17	18	19	20	21	22	23	24	25	26	27	28	29	30
80	1.022	54	52	50	48	45	43	40	37	33	30	26	22	16	9	−1
81	.056	56	54	52	49	47	44	42	39	35	32	28	24	19	13	+5
82	.091	57	55	53	51	48	46	43	40	37	34	30	27	22	17	10
83	.127	58	56	54	52	50	47	45	42	39	36	32	29	25	20	14
84	.163	60	58	56	53	51	49	46	44	41	38	35	31	27	23	18
85	1.201	61	59	57	55	53	50	48	46	43	40	37	33	30	26	21
86	.241	62	60	58	56	54	52	50	47	45	42	39	36	32	28	24
87	.281	63	62	60	58	56	54	51	49	46	44	41	38	34	31	27
88	.322	65	63	61	59	57	55	53	50	48	45	43	40	36	33	29
89	.364	66	64	62	60	58	56	54	52	50	47	44	42	38	35	31
90	1.408	67	65	64	62	60	58	56	54	51	49	46	43	40	37	34
91	.453	68	67	65	63	61	59	57	55	53	50	48	45	42	39	36
92	.499	70	68	66	64	62	61	59	57	54	52	50	47	44	41	38
93	.546	71	69	67	66	64	62	60	58	56	54	51	49	46	43	40
94	.595	72	70	69	67	65	63	61	59	57	55	53	50	48	45	42
95	1.645	73	72	70	68	66	65	63	61	59	57	54	52	50	47	44
96	.696	74	73	71	69	68	66	64	62	60	58	56	54	52	49	46
97	.749	75	74	72	71	69	67	65	64	62	60	58	55	53	51	48
98	.803	77	75	74	72	70	69	67	65	63	61	59	57	55	52	50
99	.859	78	76	75	73	72	70	68	66	64	62	60	58	56	54	52
100	1.916	79	77	76	74	73	71	69	68	66	64	62	60	58	56	53
101	1.975	80	79	77	76	74	72	71	69	67	65	63	61	59	57	55
102	2.035	81	80	78	77	75	74	72	70	68	67	65	63	61	59	56
103	.097	82	81	79	78	76	75	73	72	70	68	66	64	62	60	58
104	.160	84	82	81	79	78	76	75	74	73	71	69	68	66	64	62
105	2.225	85	83	82	80	79	77	76	74	72	71	69	67	65	63	61
106	.292	86	84	83	81	80	78	77	75	74	72	70	68	67	65	63
107	.360	87	85	84	83	81	80	78	76	75	73	72	70	68	67	64
108	.431	88	87	85	84	82	81	79	78	76	75	73	72	70	69	66
109	.503	89	88	86	85	83	82	80	79	77	76	74	73	71	69	67
110	2.576	90	89	87	86	85	83	82	80	79	77	75	74	72	70	68
111	.652	91	90	89	87	86	84	83	81	80	78	77	75	73	72	70
112	.730	92	91	90	88	87	86	84	83	81	80	78	76	75	73	71
113	.810	94	92	91	89	88	87	85	84	82	81	79	78	76	74	72
114	.891	95	93	92	91	89	88	86	85	83	82	81	80	79	77	74
115	2.975	96	94	93	92	90	90	88	88	86	85	83	82	80	78	75
116	3.061	97	96	94	93	91	90	90	89	87	86	84	83	81	80	76
117	.148	98	97	95	94	93	91	90	91	88	87	86	84	83	81	78
118	.239	99	98	96	95	94	92	91	90	88	87	85	84	82	81	79
119	.331	100	99	97	96	95	93	92	91	89	88	86	85	83	82	80
120	3.425	101	100	99	97	96	95	93	92	90	89	87	86	85	83	82
121	.522	102	101	100	98	97	96	94	93	92	90	89	87	86	84	83
122	.621	103	102	101	99	98	96	96	94	93	91	90	88	87	86	84
123	.723	104	103	102	101	99	99	98	95	94	93	91	90	88	87	85
124	.827	105	104	103	102	100	99	98	96	95	94	92	91	90	88	87
125	3.933	106	105	104	103	101	100	100	99	98	96	95	94	92	91	88
126	4.042	108	106	105	104	103	101	101	100	99	97	96	95	93	92	89
127	.154	109	107	106	105	104	102	101	100	100	98	97	96	94	93	90
128	.268	110	108	107	106	105	103	102	101	101	100	98	97	96	94	91
129	.385	111	110	108	107	106	105	103	102	102	101	99	98	97	95	93
130	4.504	112	111	109	108	107	106	105	104	103	102	101	99	98	97	94
131	.627	113	112	112	110	109	108	107	105	104	103	102	100	99	98	95
132	.752	114	113	113	111	110	109	108	107	105	104	103	101	100	99	96
133	4.880	115	114	114	113	111	110	109	108	106	105	104	103	101	100	97
134	5.011	116	115	115	114	112	111	110	109	108	106	105	104	103	101	98
135	5.145	117	116	115	113	112	111	111	110	109	107	106	105	104	102	100
136	.282	118	117	116	115	113	112	111	110	110	108	107	106	105	103	101
137	.422	119	118	117	116	114	113	113	112	111	110	108	107	106	104	102
138	.565	120	119	119	117	116	115	114	113	112	111	109	108	107	106	103
139	.712	121	120	120	119	118	117	115	114	113	112	111	110	109	107	104
140	5.862	122	121	120	119	118	116	115	114	113	112	110	109	108	107	105

TABLE A-8

Relative Humidity, Per Cent—Fahrenheit Temperatures

[Pressure = 29.0 inches]

Air Temperature t	Depression of Wet-Bulb Thermometer (t − t')																				
	0.2	0.4	0.6	0.8	1.0	1.2	1.4	1.6	1.8	2.0	2.2	2.4	2.6	2.8	3.0	3.2	3.4	3.6	3.8	4.0	4.2
−40	49																				
−39	51																				
−38	53	5																			
−37	55	9																			
−36	58	13																			
−35	61	18																			
−34	63	22																			
−33	65	26																			
−32	66	30																			
−31	68	34	3																		
−30	70	38	7																		
−29	72	42	12																		
−28	74	46	17																		
−27	76	49	22																		
−26	77	52	26	3																	
−25	78	54	30	8																	
−24	79	56	34	13																	
−23	80	58	38	18																	
−22	81	60	41	22	4																
−21	82	62	44	26	8																
−20	83	64	47	30	13																
−19	83	66	50	34	18																
−18	84	68	53	38	22	5															
−17	85	70	55	41	26	10															
−16	86	71	57	43	29	15	1														
−15	87	73	60	46	33	20	6														
−14	88	74	62	49	37	24	11														
−13	88	76	64	52	40	28	15	4													
−12	89	77	65	54	43	31	20	9													
−11	90	78	67	56	46	34	23	13	1												
−10	90	79	69	58	48	38	27	17	6												
−9	90	80	70	60	51	41	31	21	11	1											
−8	91	81	72	62	53	43	34	25	15	6											
−7	91	82	73	64	55	46	37	28	19	10	2										
−6	92	83	74	65	57	48	40	31	23	14	6										
−5	92	83	75	67	59	51	43	34	26	18	10	2									
−4	92	84	76	69	61	53	45	37	30	22	14	7									
−3	93	85	77	70	62	55	48	40	33	26	18	11	3								
−2	93	85	78	71	64	57	50	43	36	29	22	15	8	1							
−1	93	86	79	73	66	59	52	46	39	32	26	19	12	6							
0	93	87	80	74	68	61	55	48	42	35	29	23	16	10	3						
+1	94	87	81	75	69	63	57	51	44	38	32	26	20	14	8						
2	94	88	82	76	71	65	59	53	47	41	35	30	24	18	12	6	0				
3	94	89	83	77	72	66	61	55	49	44	38	33	27	21	16	10	4				
4	94	89	84	78	73	67	62	57	51	46	40	35	30	25	19	14	8	3			
5	95	89	84	79	74	69	64	58	53	48	43	38	32	27	22	17	12	7	2		
6	95	90	85	80	75	70	65	60	55	50	45	40	35	30	25	21	16	11	6	1	
7	95	90	85	80	76	71	66	61	56	52	47	43	38	33	28	24	19	14	10	5	0
8	95	91	86	81	77	72	68	63	58	54	49	45	40	35	31	27	22	18	13	9	4
9	95	91	86	82	78	73	69	64	60	55	51	47	42	38	34	29	25	21	16	12	8
10	96	91	87	83	79	74	70	66	61	57	53	49	44	40	36	32	28	24	19	15	11
11	96	92	87	83	79	75	71	67	63	59	55	51	46	42	38	34	30	26	22	18	14
12	96	92	88	84	80	76	72	68	64	60	56	52	48	45	41	37	33	29	25	21	18
13	96	92	88	85	81	77	73	69	66	62	58	54	50	47	43	39	36	32	28	24	21
14	96	92	89	85	82	78	74	71	67	63	60	56	52	49	45	42	38	34	31	27	24
15	96	93	89	86	82	79	75	72	68	65	61	58	54	51	47	44	40	37	34	30	27
16	97	93	90	86	83	79	76	73	70	66	63	59	56	53	49	46	43	40	36	33	29
17	97	93	90	87	84	80	77	74	71	67	64	61	58	54	51	48	45	42	38	35	32
18	97	94	90	87	84	81	78	75	72	68	65	62	59	56	53	50	47	44	41	38	35
19	97	94	91	88	85	82	78	75	72	69	66	63	60	57	55	52	49	46	43	40	37
20	97	94	91	88	85	82	79	76	73	70	68	65	62	59	56	53	50	47	44	42	39

(t − t')

t	4.4	4.6	4.8	5.0	5.2	5.4	5.6	5.8	6.0	6.2	6.4	6.6
9	4											
10	7	3										
11	11	7	3									
12	14	10	6	3								
13	17	14	10	6	3							
14	20	17	13	10	6	3						
15	23	20	17	13	10	6	3					
16	26	23	20	16	13	10	7	3	0			
17	29	26	23	19	16	13	10	7	4	1		
18	32	29	25	22	19	16	13	10	7	4	1	
19	34	31	28	25	22	19	16	13	11	8	5	2
20	36	33	30	28	25	22	19	16	14	11	8	5

(t − t')

t	0.1	0.2	0.3	0.4	0.5
−50	50				
−49	54	9			
−48	57	15			
−47	60	20			
−46	63	25			
−45	66	30			
−44	68	35			
−43	70	39	6		
−42	71	43	11		
−41	72	46	16		
−40	73	49	20		
−39	74	51	24		
−38	75	53	27	5	
−37	76	55	31	9	
−36	77	58	35	13	
−35	78	61	39	18	
−34	80	63	43	22	4
−33	81	65	46	26	9
−32	82	66	48	30	13
−31	83	68	51	34	18
−30	84	70	54	38	23

TABLE A–8 *(Continued)*

Relative Humidity, Per Cent—Fahrenheit Temperatures

[Pressure = 29.0 inches]

Air Temperature t	Depression of Wet-Bulb Thermometer $(t - t')$																				
	0.5	1.0	1.5	2.0	2.5	3.0	3.5	4.0	4.5	5.0	5.5	6.0	6.5	7.0	7.5	8.0	8.5	9.0	9.5	10.0	10.5
20	92	85	78	70	63	56	49	42	35	28	21	14	7								
21	93	86	78	71	64	57	50	44	37	30	24	17	10	3							
22	93	86	79	72	65	59	52	45	39	32	26	19	13	7	0						
23	93	87	80	73	66	60	53	47	41	34	28	22	16	10	3						
24	94	87	81	74	68	61	55	49	42	36	30	24	18	12	6	0					
25	94	87	81	75	69	63	56	50	44	38	32	27	21	15	9	4					
26	94	88	82	75	69	64	58	52	46	40	34	29	23	18	12	7	1				
27	94	88	82	76	70	65	59	53	48	42	36	31	26	20	15	9	4				
28	94	88	82	77	71	66	60	55	49	44	38	33	28	23	17	12	7	2			
29	94	89	83	78	72	67	61	56	51	45	40	35	30	25	20	15	10	5	0		
30	95	89	84	78	73	68	62	57	52	47	42	37	32	27	22	17	12	8	3		
31	95	89	84	79	74	69	63	58	53	49	44	39	34	29	24	20	15	10	6	1	
32	95	90	85	79	74	69	65	60	55	50	45	41	36	31	26	22	17	13	9	4	
33	95	90	85	80	76	71	66	61	56	52	47	42	38	33	29	24	20	16	11	7	3
34	95	90	86	81	77	72	67	62	58	53	49	44	40	35	31	27	22	18	14	9	5
35	95	91	86	82	77	73	68	64	59	55	50	46	41	37	33	29	24	20	16	12	8
36	95	91	87	82	78	73	69	65	61	56	52	48	43	39	35	31	27	23	18	14	10
37	95	91	87	83	79	74	70	66	62	58	54	49	45	41	37	33	29	25	21	17	13
38	96	91	87	83	79	75	71	67	63	59	55	51	47	43	39	35	31	27	23	19	15
39	96	92	88	84	80	76	72	68	64	60	56	52	48	44	41	37	33	29	25	21	17
40	96	92	88	84	80	76	72	68	64	61	57	53	49	46	42	38	35	31	27	23	20
41	96	92	88	84	80	77	73	69	65	62	58	54	50	47	43	40	36	33	29	26	22
42	96	92	88	85	81	77	73	70	66	62	59	55	51	48	45	41	38	34	31	28	24
43	96	92	88	85	81	78	74	70	67	63	60	56	52	49	46	43	39	36	32	29	26
44	96	93	89	85	82	78	74	71	68	64	61	57	54	51	47	44	40	37	34	31	28
45	96	93	89	86	82	79	75	71	68	65	61	58	55	52	48	45	42	39	36	33	29
46	96	93	89	86	82	79	75	72	69	65	62	59	56	53	49	46	43	40	37	34	31
47	96	93	89	86	83	79	76	73	69	66	63	60	57	54	50	47	44	41	38	35	32
48	96	93	90	87	83	80	76	73	70	67	63	60	57	54	51	48	45	42	39	36	34
49	96	93	90	87	83	80	77	74	71	67	64	61	58	55	52	49	46	43	40	37	35
50	96	93	90	87	84	81	77	74	71	68	65	62	59	56	53	50	47	44	42	39	36
51	97	94	90	87	84	81	78	75	72	69	66	63	60	57	54	51	48	45	43	40	37
52	97	94	91	88	84	81	78	75	72	69	66	63	60	58	55	52	49	46	44	41	39
53	97	94	91	88	85	82	78	75	73	70	67	64	61	58	56	53	50	47	45	42	40
54	97	94	91	88	85	82	79	76	73	70	67	65	62	59	57	54	51	48	46	43	41
55	97	94	91	88	85	82	79	76	74	71	68	65	62	60	57	55	52	49	47	44	42
56	97	94	91	88	85	82	79	77	74	71	69	66	63	61	58	55	53	50	48	45	43
57	97	94	91	88	85	83	80	77	74	72	69	66	64	61	59	56	53	51	49	46	44
58	97	94	91	89	86	83	80	77	75	72	69	67	64	62	60	57	54	52	49	47	45
59	97	94	92	89	86	83	81	78	75	73	70	68	65	63	60	58	55	53	50	48	45
60	97	94	92	89	86	84	81	78	76	73	71	68	65	63	61	58	56	53	51	49	46
61	97	94	92	89	86	84	81	79	76	74	71	68	66	64	61	59	56	54	52	50	47
62	97	94	92	89	87	84	81	79	77	74	72	69	66	64	62	60	57	55	53	50	48
63	97	95	92	90	87	84	82	79	77	74	72	70	67	65	62	60	58	56	53	50	49
64	97	95	92	90	87	85	82	79	77	75	72	70	68	66	63	61	58	56	54	52	50
65	97	95	92	90	87	85	82	80	78	75	73	70	68	66	64	62	59	57	55	53	50
66	97	95	92	90	87	85	83	80	78	76	73	71	68	66	64	62	60	58	56	54	51
67	97	95	92	90	88	85	83	80	78	76	74	71	69	67	65	62	60	58	56	54	52
68	97	95	93	90	88	85	83	81	78	76	74	72	69	67	65	63	61	59	57	55	53
69	97	95	93	90	88	86	83	81	79	77	74	72	70	68	66	64	61	59	57	55	53
70	98	95	93	90	88	86	83	81	79	77	75	72	70	68	66	64	62	60	58	56	54
71	98	95	93	90	88	86	84	82	79	77	75	73	71	69	67	65	62	60	59	56	54
72	98	95	93	91	89	86	84	82	80	78	75	73	71	69	67	65	63	61	59	57	55
73	98	95	93	91	89	86	84	82	80	78	76	73	71	69	67	66	63	61	60	58	56
74	98	95	93	91	89	86	84	82	80	78	76	74	72	70	68	66	64	62	60	58	56
75	98	96	93	91	89	87	84	82	80	78	76	74	72	70	68	66	64	63	61	59	57
76	98	96	93	91	89	87	85	83	80	78	76	74	72	70	68	67	65	63	61	59	57
77	98	96	93	91	89	87	85	83	81	79	77	75	73	71	69	67	65	63	62	60	58
78	98	96	94	91	89	87	85	83	81	79	77	75	73	71	69	67	66	64	62	60	58
79	98	96	94	91	89	87	85	83	81	79	77	75	73	71	70	68	66	64	62	60	59
80	98	96	94	91	89	87	85	83	81	79	77	76	74	72	70	68	66	64	63	61	59

TABLE A–8 (Continued)

Relative Humidity, Per Cent—Fahrenheit Temperatures

[Pressure = 29.0 inches]

Air Temperature t	Depression of Wet-Bulb Thermometer (t − t')																				
	11.0	11.5	12.0	12.5	13.0	13.5	14.0	14.5	15.0	15.5	16.0	16.5	17.0	17.5	18.0	18.5	19.0	19.5	20.0	20.5	21.0
34	1																				
35	4																				
36	6	3																			
37	9	5	1																		
38	12	8	4	0																	
39	14	10	7	3																	
40	16	13	9	6	2																
41	18	15	11	8	5	1															
42	21	17	14	10	7	4	0														
43	23	19	16	13	9	6	3														
44	24	21	18	15	12	9	5	2													
45	26	23	20	17	14	11	8	5	2												
46	28	25	22	19	16	13	10	7	4	1											
47	29	26	23	20	17	15	12	9	6	3	1										
48	31	28	25	22	19	16	14	11	8	6	3	0									
49	32	29	26	24	21	18	15	13	10	7	5	2									
50	33	31	28	25	22	20	17	14	12	9	7	4	2								
51	35	32	29	27	24	21	19	16	14	11	9	6	4	1							
52	36	33	30	28	25	23	20	18	15	13	10	8	6	3	0						
53	37	34	32	29	27	24	22	19	17	15	12	10	7	5	3	0					
54	38	35	33	30	28	26	23	21	18	16	14	12	9	7	5	2	0				
55	39	37	34	32	29	27	25	22	20	18	15	13	11	9	6	4	2				
56	40	38	35	33	31	28	26	24	21	19	17	15	12	10	8	6	4	2			
57	41	39	36	34	32	29	27	25	23	20	18	16	14	12	10	7	5	3	1		
58	42	40	38	35	33	31	28	26	24	22	20	17	15	13	11	9	7	5	3	1	
59	43	41	39	36	34	32	30	27	25	23	21	19	17	15	13	11	9	7	5	3	1
60	44	42	40	37	35	33	31	29	27	25	22	20	18	16	14	12	10	8	6	4	2
61	45	43	40	38	36	34	32	30	28	26	24	22	20	18	16	14	12	10	8	6	4
62	46	44	41	39	37	35	33	31	29	27	25	23	21	19	17	15	13	11	9	8	6
63	47	45	42	40	38	36	34	32	30	28	26	24	22	20	18	16	14	13	11	9	7
64	48	45	43	41	39	37	35	33	31	29	27	25	23	22	20	18	16	14	12	11	9
65	48	46	44	42	40	38	36	34	32	30	28	26	25	23	21	19	17	15	13	12	10
66	49	47	45	43	41	39	37	35	33	31	29	27	26	24	22	20	18	17	15	13	11
67	50	48	46	44	42	40	38	36	34	32	30	29	27	25	23	21	20	18	16	15	13
68	51	49	47	45	43	41	39	37	35	33	31	30	28	26	24	23	21	19	17	16	14
69	51	49	47	45	44	42	40	38	36	34	32	31	29	27	25	24	22	20	19	17	15
70	52	50	48	46	44	42	40	39	37	35	33	32	30	28	26	25	23	21	20	18	17
71	53	51	49	47	45	43	41	39	38	36	34	32	31	29	27	26	24	22	21	19	18
72	53	51	49	48	46	44	42	40	39	37	35	33	32	30	28	27	25	23	22	20	19
73	54	52	50	48	46	45	43	41	40	38	36	34	33	31	29	28	26	24	23	21	20
74	54	53	51	49	47	45	44	42	40	39	37	35	34	32	30	29	27	25	24	22	21
75	55	53	51	50	48	46	44	43	41	39	38	36	34	33	31	30	28	26	25	23	22
76	55	54	52	50	48	47	45	43	42	40	38	37	35	34	32	30	29	27	26	24	23
77	56	54	52	51	49	47	46	44	42	41	39	37	36	34	33	31	30	28	27	25	24
78	57	55	53	51	50	48	46	45	43	42	40	38	37	35	34	32	31	29	28	26	25
79	57	55	54	52	50	49	47	46	44	43	41	39	37	36	34	33	31	30	29	27	26
80	57	56	54	52	51	49	47	46	44	43	41	40	38	37	35	34	32	31	29	28	27

t	Depression of Web-Bulb Thermometer (t − t')																			
	21.5	22.0	22.5	23.0	23.5	24.0	24.5	25.0	25.5	26.0	26.5	27.0	27.5	28.0	28.5	29.0	29.5	30.0	30.5	31.0
60	1																			
61	2	0																		
62	4	2	0																	
63	5	4	2	0																
64	7	5	3	2																
65	8	7	5	3	2	1														
66	10	8	6	5	3	1														
67	11	9	8	6	5	3	1													
68	12	11	9	8	6	4	3	1												
69	14	12	10	9	7	6	4	3	1											
70	15	13	12	10	9	7	6	4	3	1										
71	16	15	13	11	10	8	7	5	4	3	1									
72	17	16	14	13	11	10	8	7	5	4	2	1								
73	18	17	15	14	12	11	9	8	7	5	4	2	1							
74	19	18	16	15	14	12	11	9	8	7	5	4	2	1						
75	20	19	17	16	15	13	12	11	9	8	6	5	4	2	1					
76	21	20	19	17	16	14	13	12	10	9	8	6	5	4	2	1				
77	22	21	20	18	17	15	14	13	11	10	9	7	6	5	4	2	1			
78	23	22	21	19	18	16	15	14	12	11	10	9	7	6	5	4	2	1	0	
79	24	23	21	20	19	17	16	15	13	12	11	10	8	7	6	5	4	2	1	0
80	25	24	22	21	20	18	17	16	14	13	12	11	10	9	8	7	6	5	4	2

TABLE A–8 (*Continued*)
Relative Humidity, Per Cent—Fahrenheit Temperatures
[Pressure = 29.0 inches]

Air Temperature t	Depression of Wet-Bulb Thermometer (t − t′)														
	1	2	3	4	5	6	7	8	9	10	11	12	13	14	15
80	96	91	87	83	79	76	72	68	64	61	57	54	51	47	44
82	96	92	88	84	80	76	72	69	65	62	58	55	52	49	46
84	96	92	88	84	80	77	73	70	66	63	59	56	53	50	47
86	96	92	88	85	81	77	74	70	67	63	60	57	54	51	48
88	96	92	88	85	81	78	74	71	67	64	61	58	55	52	49
90	96	92	89	85	81	78	75	71	68	65	62	59	56	53	50
92	96	92	89	85	82	78	75	72	69	65	62	59	57	54	51
94	96	93	89	86	82	79	75	72	69	66	63	60	57	54	52
96	96	93	89	86	82	79	76	73	70	67	64	61	58	55	53
98	96	93	89	86	83	79	76	73	70	67	64	61	59	56	53
100	96	93	90	86	83	80	77	74	71	68	65	62	59	57	54
102	96	93	90	86	83	80	77	74	71	68	65	63	60	57	55
104	97	93	90	87	84	80	77	74	72	69	66	63	61	58	56
106	97	93	90	87	84	81	78	75	72	69	66	64	61	59	56
108	97	93	90	87	84	81	78	75	72	70	67	64	62	59	57
110	97	94	90	87	84	81	78	76	73	70	67	65	62	60	57
112	97	94	90	87	84	82	79	76	73	70	68	65	63	60	58
114	97	94	91	88	85	82	79	76	74	71	68	66	63	61	59
116	97	94	91	88	85	82	79	77	74	71	69	66	64	61	59
118	97	94	91	88	85	82	79	77	74	72	69	67	64	62	60
120	97	94	91	88	85	82	80	77	74	72	69	67	65	62	60
122	97	94	91	88	85	83	80	77	75	72	70	67	65	63	61
124	97	94	91	88	86	83	80	78	75	73	70	68	65	63	61
126	97	94	91	89	86	83	81	78	75	73	71	68	66	64	62
128	97	94	91	89	86	83	81	78	76	73	71	69	66	64	62
130	97	94	92	89	86	84	81	78	76	74	71	69	67	65	62
132	97	94	92	89	86	84	81	79	76	74	72	69	67	65	63
134	97	94	92	89	86	84	81	79	76	74	72	70	67	65	63
136	97	94	92	89	87	84	82	79	77	74	72	70	68	66	64
138	97	94	92	89	87	84	82	79	77	75	72	70	68	66	64
140	97	95	92	89	87	84	82	80	77	75	73	71	68	66	64

t	Depression of Wet-Bulb Thermometer (t − t′)														
	16	17	18	19	20	21	22	23	24	25	26	27	28	29	30
80	41	38	35	32	29	27	24	21	18	16	13	11	8	6	4
82	43	40	37	34	31	28	25	23	20	18	15	13	10	8	6
84	44	41	38	35	32	30	27	25	22	20	17	15	12	10	8
86	45	42	39	37	34	31	29	26	24	21	19	17	14	12	10
88	46	43	41	38	35	33	30	28	25	23	21	18	16	14	12
90	47	44	42	39	37	34	32	29	27	24	22	20	18	16	14
92	48	45	43	40	38	35	33	30	28	26	24	22	19	17	15
94	49	46	44	41	39	36	34	32	29	27	25	23	21	19	17
96	50	47	45	42	40	37	35	33	31	29	26	24	22	20	18
98	51	48	46	43	41	39	36	34	32	30	28	26	24	22	20
100	52	49	47	44	42	40	37	35	33	31	29	27	25	23	21
102	52	50	47	45	43	41	38	36	34	32	30	28	26	24	22
104	53	51	48	46	44	41	39	37	35	33	31	29	27	25	24
106	54	51	49	47	45	42	40	38	36	34	32	30	28	27	25
108	54	52	50	47	45	43	41	39	37	35	33	31	29	28	26
110	55	53	50	48	46	44	42	40	38	36	34	32	30	29	27
112	56	53	51	49	47	45	43	41	39	37	35	33	31	30	28
114	56	54	52	50	48	45	43	41	40	38	36	34	32	31	29
116	57	55	52	50	48	46	44	42	40	38	37	35	33	31	30
118	57	55	53	51	49	47	45	43	41	39	37	36	34	32	31
120	58	56	54	51	49	47	46	44	42	40	38	36	35	33	31
122	58	56	54	52	50	48	46	44	42	41	39	37	36	34	32
124	59	57	55	53	51	49	47	45	43	41	40	38	36	35	33
126	59	57	55	53	51	49	47	46	44	42	40	39	37	35	34
128	60	58	56	54	52	50	48	46	44	43	41	39	38	36	34
130	60	58	56	54	52	50	49	47	45	43	42	40	38	37	35
132	61	59	57	55	53	51	49	47	46	44	42	41	39	37	36
134	61	59	57	55	53	51	50	48	46	44	43	41	40	38	36
136	61	59	58	56	54	52	50	48	47	45	43	42	40	39	37
138	62	60	58	56	54	52	51	49	47	45	44	42	41	39	38
140	62	60	58	56	55	53	51	49	48	46	44	43	41	40	38

TABLE A–8 *(Continued)*
Relative Humidity, Per Cent—Fahrenheit Temperatures
[Pressure = 29.0 inches]

Air Temperature t	Depression of Wet-Bulb Thermometer (t − t′)														
	1	2	3	4	5	6	7	8	9	10	11	12	13	14	15
80	96	91	87	83	79	76	72	68	64	61	57	54	51	47	44
82	96	92	88	84	80	76	72	69	65	62	58	55	52	49	46
84	96	92	88	84	80	77	73	70	66	63	59	56	53	50	47
86	96	92	88	85	81	77	74	70	67	63	60	57	54	51	48
88	96	92	88	85	81	78	74	71	67	64	61	58	55	52	49
90	96	92	89	85	81	78	75	71	68	65	62	59	56	53	50
92	96	92	89	85	82	78	75	72	69	65	62	59	57	54	51
94	96	93	89	86	82	79	75	72	69	66	63	60	57	54	52
96	96	93	89	86	82	79	76	73	70	67	64	61	58	55	53
98	96	93	89	86	83	79	76	73	70	67	64	61	59	56	53
100	96	93	90	86	83	80	77	74	71	68	65	62	59	57	54
102	96	93	90	86	83	80	77	74	71	68	65	63	60	57	55
104	97	93	90	87	84	80	77	74	72	69	66	63	61	58	56
106	97	93	90	87	84	81	78	75	72	69	66	64	61	59	56
108	97	93	90	87	84	81	78	75	72	70	67	64	62	59	57
110	97	94	90	87	84	81	78	76	73	70	67	65	62	60	57
112	97	94	90	87	84	82	79	76	73	70	68	65	63	60	58
114	97	94	91	88	85	82	79	76	74	71	68	66	63	61	59
116	97	94	91	88	85	82	79	77	74	71	69	66	64	61	59
118	97	94	91	88	85	82	79	77	74	72	69	67	64	62	60
120	97	94	91	88	85	82	80	77	74	72	69	67	65	62	60
122	97	94	91	88	85	83	80	77	75	72	70	67	65	63	61
124	97	94	91	88	86	83	80	78	75	73	70	68	65	63	61
126	97	94	91	89	86	83	81	78	75	73	71	68	66	64	62
128	97	94	91	89	86	83	81	78	76	73	71	69	66	64	62
130	97	94	92	89	86	84	81	78	76	74	71	69	67	65	62
132	97	94	92	89	86	84	81	79	76	74	72	69	67	65	63
134	97	94	92	89	86	84	81	79	76	74	72	70	67	65	63
136	97	94	92	89	87	84	82	79	77	74	72	70	68	66	64
138	97	94	92	89	87	84	82	79	77	75	72	70	68	66	64
140	97	95	92	89	87	84	82	80	77	75	73	71	68	66	64

t	Depression of Wet-Bulb Thermometer (t − t′)														
	16	17	18	19	20	21	22	23	24	25	26	27	28	29	30
80	41	38	35	32	29	27	24	21	18	16	13	11	8	6	4
82	43	40	37	34	31	28	25	23	20	18	15	13	10	8	6
84	44	41	38	35	32	30	27	25	22	20	17	15	12	10	8
86	45	42	39	37	34	31	29	26	24	21	19	17	14	12	10
88	46	43	41	38	35	33	30	28	25	23	21	19	16	14	12
90	47	44	42	39	37	34	32	29	27	24	22	20	18	16	14
92	48	45	43	40	38	35	33	30	28	26	24	22	19	17	15
94	49	46	44	41	39	36	34	32	29	27	25	23	21	19	17
96	50	47	45	42	40	37	35	33	31	29	26	24	22	20	18
98	51	48	46	43	41	39	36	34	32	30	28	26	24	22	20
100	52	49	47	44	42	40	37	35	33	31	29	27	25	23	21
102	52	50	47	45	43	41	38	36	34	32	30	28	26	24	22
104	53	51	48	46	44	42	39	37	35	33	31	29	27	25	24
106	54	51	49	47	45	42	40	38	36	34	32	30	28	26	25
108	54	52	50	47	45	43	41	39	37	35	33	31	29	28	26
110	55	53	50	48	46	44	42	40	38	36	34	32	30	28	27
112	56	53	51	49	47	45	43	41	39	37	35	33	31	30	28
114	56	54	52	50	48	45	43	41	40	38	36	34	32	31	29
116	57	55	52	50	48	46	44	42	40	38	37	35	33	31	30
118	57	55	53	51	49	47	45	43	41	39	37	36	34	32	31
120	58	56	54	51	49	47	46	44	42	40	38	36	35	33	31
122	58	56	54	52	50	48	46	44	42	41	39	37	36	34	32
124	59	57	55	53	51	49	47	45	43	41	40	38	36	35	33
126	59	57	55	53	51	49	47	46	44	42	40	39	37	35	34
128	60	58	56	54	52	50	48	46	44	42	41	39	38	36	34
130	60	58	56	54	52	50	49	47	45	43	42	40	38	37	35
132	61	59	57	55	53	51	49	47	46	44	42	41	39	38	36
134	61	59	57	55	53	51	50	48	46	44	43	41	40	38	36
136	61	59	58	56	54	52	50	48	47	45	43	42	40	39	37
138	62	60	58	56	54	52	51	49	47	45	44	42	41	39	38
140	62	60	58	56	55	53	51	49	48	46	44	43	41	40	38

REDUCTION OF PRESSURE TO SEA LEVEL

The United States Weather Bureau has prepared a "Pressure-Altitude Chart (1050–800 mb.)," W. B. Form 1154A, which may be used to reduce *station pressure* to *sea-level pressure*. This chart together with W. B. Form 1154B, "Humidity Correction to Mean Temperature for Pressure Reduction," may be used if individual pressure reduction tables are not available at the observatory.

To reduce *station pressure* to sea level:

1. Determine Hb = the station elevation.
2. Express Hb in terms of the unit: 0.98 geodynamic meter (gdm), by conversion from geometric meters.

$$Hb \text{ (in 0.98 gdm)} = F \times Hb \text{ (in geometric meters)}$$

where F is obtained from Table A–9, using the latitude of the observatory.

TABLE A–9

Latitude	F
90	1.0033
80	1.0031
70	1.0027
60	1.0020
50	1.0011
40	1.0002
30	0.9993
20	0.9986
10	0.9982
0	0.9980

3. Determine P = "station pressure" in millibars.
4. Determine t_{mv} = the mean virtual temperature of the fictitious air column assumed to extend from the station to sea level as follows:

t_m = mean (dry-bulb) temperature assuming a lapse rate of 1° C. per 200 meters or $5\frac{1}{2}$° F. per 2,000 feet.

t_s = the temperature argument (mean temperature for last 12 hours).

$t_m = t_s \left(\dfrac{Hb \times 1.8}{400} \right)$, where t_m and t_s are °F. and Hb is in 0.98 gdm, or

$t_m = t_s \left(\dfrac{Hb}{400} \right)$, where t_m and t_s are °C. and Hb is in 0.98 gdm.

Next apply the correction for humidity as follows:

$$t_{mv} = t_m + C_m$$

where C_m is obtained from the humidity correction chart.

5. By means of the elevation scale on the left of the pressure-altitude chart, mark off on a straight edge the distance corresponding to Hb from 0 elevation.

6. Apply the straight edge to the chart so that it coincides with the vertical line corresponding to t_{mv}, the mean virtual temperature obtained in paragraph 4 above. Then slide the straight edge vertically until the mark Hb is on the sloping pressure line for the "station pressure."

7. Read the reduced, sea-level pressure on the sloping line which crosses the 0 or sea-level elevation on the straight edge.

EXAMPLE

Station elevation, Hb = 750 (0.98 gdm)
Station pressure, P = 938.0 millibars
Station temperature argument, t_s = 50° F.
Dew point, d.p.   = 32° F.

$$t_m = t_s + \left(\frac{Hb \times 1.8}{400}\right) = 50° \text{ F.} + 3.4° \text{ F.} = 53.4° \text{ F.}$$

From humidity correction chart
Hb = 750 and dew point 32° F.
C_m = 1.4° F.
$t_{mv} = t_m + C_m = 53.4°$ F. + 1.4° F. = 54.8° F.

Using the straight edge the reduced sea-level pressure = 1026.0 mb.

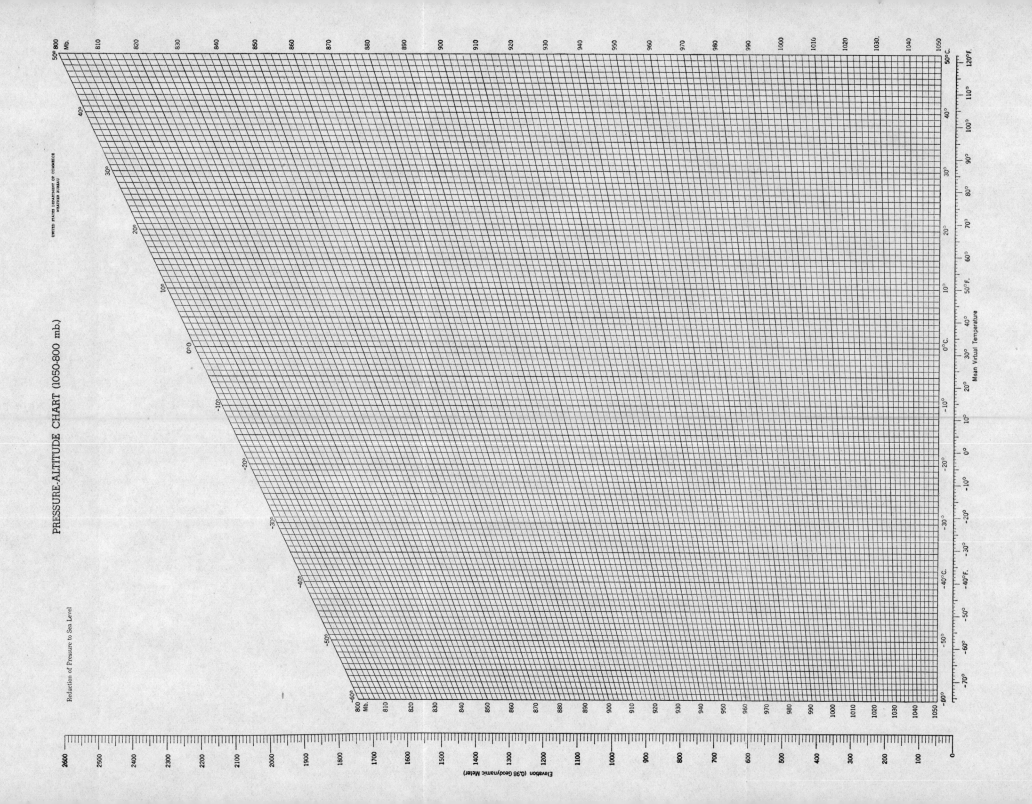

PRESSURE-ALTITUDE CHART (1050-800 mb.)

UNITED STATES DEPARTMENT OF COMMERCE
WEATHER BUREAU

Reduction of Pressure to Sea Level

Mean Virtual Temperature

Elevation (0.98 Geodynamic Meter)

HUMIDITY CORRECTION TO MEAN TEMPERATURE FOR PRESSURE REDUCTION

This chart is primarily intended for use in conjunction with WB Form 1154A, "Pressure-Altitude Chart (1050-800 mb.)", which is designed to permit reduction of pressure to sea level. The slanting lines hereon represent values of C_m = "Correction for Mean Moisture Content of Air Column." The arguments for obtaining C_m are dew point at time of observation (horizontal scale) and station elevation, H_b, in 0.98 gdm. (vertical scale). To facilitate the reading of C_m from this chart, draw a horizontal line to correspond to H_b in accordance with the elevation scale at the sides.

The application of C_m is seen from the relationship:

$$t_{mv} = t_m + C_m, \text{ where}$$

t_{mv} = mean virtual temperature of the air column extending from the station down to sea level, and

t_m = mean (dry-bulb) temperature of that air column. Values of t_{mv} are required to reduce pressure to sea level, and serve as the argument of the horizontal scale on WB Form 1154A.

CAUTION: *Data printed in slanting type (italics) refer to degrees Centigrade, and data in vertical type refer to* degrees Fahrenheit. *Consistency of units (°C. or °F.) is essential for t_m, C_m, t_{mv}, etc.*

BASIS OF CHART: C_m has been computed on the basis of the assumptions that barometric pressure varies with height in accordance with the U. S. Standard Atmosphere, and that vapor pressure varies with height in accordance with Hann's equation:

$$e/e_o = 10^{-h/6300}$$

, where e_o = vapor pressure at height 0, and e = vapor pressure at height h, in meters.

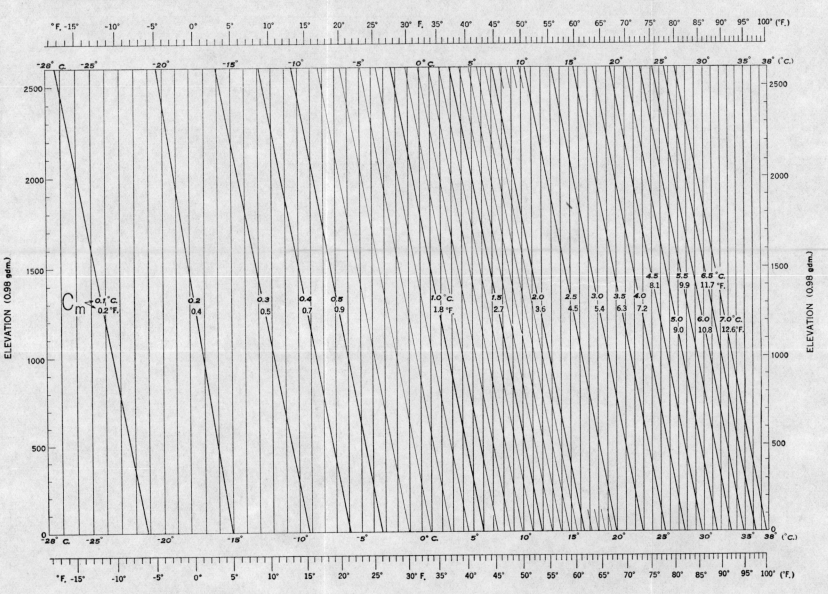

WORK SHEET A

Convert the following values from Centigrade to Fahrenheit and from Fahrenheit to Centigrade.

TEMPERATURE CENTIGRADE	TEMPERATURE FAHRENHEIT
0 degrees	_____ degrees
32 degrees	_____ degrees
−32 degrees	_____ degrees
100 degrees	_____ degrees

TEMPERATURE FAHRENHEIT	TEMPERATURE CENTIGRADE
0 degrees	_____ degrees
32 degrees	_____ degrees
212 degrees	_____ degrees
79 degrees	_____ degrees
54 degrees	_____ degrees
27 degrees	_____ degrees
−7 degrees	_____ degrees
−29 degrees	_____ degrees

14.7 pounds per square inch equals _____ inches or equals also _____ millibars of pressure.

Appendix

WORK SHEET B

Single Interpolations

Using the psychrometric tables, compute the actual dew point and relative humidity for the values indicated herein.

Number	Dry-Bulb Temperature	Wet-Bulb Temperature	Dew Point	Relative Humidity	Dew Point Recorded	Relative Humidity Recorded
1	70.0	60.0				
2	73.0	67.0				
3	72.0	65.5				
4	55.0	46.5				
5	51.0	46.1				
6	40.0	36.2				
7	39.0	38.3				
8	24.0	19.4				
9	22.0	20.6				
10	21.0	20.7				
11	29.0	26.8				
12	33.0	30.9				
13	80.0	70.0				
14	81.0	75.5				
15	82.0	78.5				
16	88.0	79.1				
17	91.0	70.2				
18	92.0	75.3				
19	101.0	79.6				
20	99.0	77.4				

WORK SHEET C

Double Interpolations

Using the psychrometric tables, compute the actual dew points and relative humidities for the values listed herein.

Number	Dry-Bulb Temperature	Wet-Bulb Temperature	Actual Dew Point	Relative Humidity	Dew Point Recorded	Relative Humidity Recorded
1	100.4	80.6				
2	45.8	43.1				
3	11.6	9.4				
4	1.8	1.6				
5	81.7	70.5				
6	75.6	66.2				
7	32.3	25.5				
8	72.9	72.1				
9	1.2	0.5				
10	40.3	32.2				
11	50.2	48.3				
12	0.6	0.2				
13	75.2	50.6				
14	20.3	18.3				
15	−12.5	−12.7				
16	90.4	60.2				
17	26.3	20.2				
18	103.2	70.6				
19	55.4	53.6				
20	70.8	60.2				

WORK SHEET D

Computation of Sea-Level Pressures in Inches and Millibars

Compute the sea-level pressures in inches and millibars from the data indicated herein. Station elevation $Hb = 1500$ feet.

Time	Temper- ature	Time	Temper- ature	Station Pressure	Sea-Level Pressure (in.)	Sea-Level Pressure (mb.)
1930	44.2	0830	77.2	28.11		
2030	25.5	0930	44.3	28.44		
2130	50.4	1030	21.1	28.79		
2230	18.2	1130	14.2	28.32		
2330	10.4	1230	−0.2	28.63		
0030	1.5	1330	−11.5	28.85		
0130	−10.0	1430	−14.4	29.07		
0230	0.0	1530	42.5	28.25		
0330	30.1	1630	60.1	28.93		
0430	35.2	1730	100.2	28.47		
0530	90.1	1830	17.1	28.81		
0630	14.4	1930	14.3	28.48		
0730	20.2	2030	33.4	28.36		
0830	44.5	2130	91.2	28.67		
0930	80.2	2230	72.4	28.22		
1030	61.1	2330	44.1	29.09		
1130	50.7	0030	88.3	28.74		
1230	77.1	0130	18.4	28.56		
1330	44.2	0230	0.3	28.94		
1430	−5.1	0330	−1.2	28.17		
1530	16.4	0430	41.9	29.03		
1630	11.2	0530	99.0	29.58		
1730	80.2	0630	11.4	28.29		
1830	−5.1	0730	12.2	28.55		
1930	−0.2					

WORK SHEET E

Computation of Sea-Level Pressures

Compute sea-level pressure for station, sea-level pressure in inches and millibars from the data indicated herein. Station elevation Hb = 1050 feet.

Number	Attached Thermometer	Reading Observed	Station Pressure	Current Temperature	12-Hour Temperature	Sea Level in.	Sea Level mb.
1	77.5	29.000		54.2	77.5		
2	69.5	28.500		59.4	83.3		
3	72.5	28.100		60.1	44.1		
4	76.0	28.624		−11.0	−5.1		
5	69.5	28.352		80.2	90.2		
6	80.0	28.584		70.4	71.4		
7	71.0	28.243		56.6	56.6		
8	75.5	29.072		100.2	60.5		
9	82.0	29.111		9.5	8.4		
10	71.5	28.241		14.4	7.2		
11	87.5	28.221		4.3	−2.4		
12	88.0	29.083		55.1	41.1		
13	84.5	28.942		22.4	17.3		
14	82.0	28.241		60.5	50.2		
15	68.0	28.641		10.2	5.1		
16	66.5	28.420		0.0	7.2		
17	74.0	28.999		52.3	44.4		
18	71.5	29.092		−21.2	−10.4		
19	77.0	28.891		50.1	63.3		

WORK SHEET F

Complete this form by filling in the spaces shown herein with the correct information.

Name	Type	Appearance	Composition	How Distinguished from Other
Ci				Ci from Cs
Cs				Cs from As
Cc				Cc from Ac
Ac				Ac from Sc
As				As from Ns
Ns				Ns from St
Sc				Sc from Cu
Cu				Cu from Cb
Cb				Cb from Ns
St				St from Sc
Fs				Fs from St
Fc				Fc from Cu
Acc				Acc from Ac
Mc				Mc from Cb

WORK SHEET G

Observing and Typing Clouds

By observing and typing the clouds at given times during the day, complete form by posting the data required.

Observation Number	Time	Low Cloud/s	Middle Cloud/s	High Cloud/s	Vertical Type Cloud/s
1					
2					
3					
4					
5					
6					
7					
8					
9					
10					
11					
12					
13					
14					
15					
16					
17					
18					
19					
20					
21					
22					
23					
24					

WORK SHEET H

Identification of Cloud Types and Obtaining Characteristic Levels

Complete this form by filling in the spaces herein by the identification of cloud types and ranges of heights for clouds listed below.

Types	Identification of Types	Ranges of Heights	
		High	Low
Ci			
Cs			
Cc			
Ac			
As			
Ns			
Sc			
Cu			
Cb			
St			

WORK SHEET I

Estimate Ceiling Heights

Complete this form by filling in the spaces with the pertinent information for estimating cloud height.

Cloud Type	Range	Usual Height	Seasonal Variation	Locality Variation	Appearance	
					If lower than usual	If higher than usual
Ci						
Cs						
Cc						
Ac						
As						
Ns						
Sc						
Cu						
Cb						
St						

QUESTIONS

1. Name three visual weather observation elements.

   Ans.   *a.* _____

         b. _____

         c. _____

2. Name four instrumental weather observation elements.

   Ans. *a.* _____

   b. _____

   c. _____

   d. _____

3. Express atmospheric pressures in the units given below:

   a. 29.92 inches of mercury _____ millibars

   b. _____ pounds per square inch

4. Why use coded reports?

   Ans. _____

   _____

5. What interests does climatology serve?

   Ans. _____

6. The degree of accuracy required to read Weather Bureau thermometers is which of the following:

   a. Whole degrees   *d.* Quarter degrees

   b. Half degrees   *e.* Tenths of degrees

   c. Hundredths of degrees

7. Why are corrections, plus or minus, used with observed thermometer readings?

   Ans. _____

   _____

8. Using the psychrometric tables, obtain and enter dew points for the following:

   a. Dry bulb   45.8   *c.* Dry bulb   −11.4
   　　 Wet bulb   43.1   　　 Wet bulb   −11.7
   　　　　　　 ____   ____

   　　　　　　　　 Ans.   *Ans.*

   b. Dry bulb   0.8
   　　 Wet bulb   −0.3
   　　　　　　 ____

   　　　　　　　　 Ans.

9. Three types of pressure-measuring instruments may be used. One is a _____ the second is an _____ barometer and the third is called a _____.

10. The most accurate of the three types of barometers is the _____ type.

11. Why does the Weather Bureau use three types of barometers?

12. The _____ barometer is used to correct a _____.

13. Does a station barometer read sea-level pressure in this region?

14. Before a barometric reading can be used to correct other pressure recording instruments a temperature _____ table must be used.

15. Barometer reads 28.35 at your station. The attached thermometer reads 72.5. What is the station pressure?

16. By what means and how frequently do you correct barograph readings?

    Ans. _____

    _____

17. What method or methods may be used in estimating wind direction?

    Ans. _____

    _____

18. Name the cardinal points of the compass.

    Ans. _____

    _____

19. What exposure is assumed in the Beaufort wind velocity equivalent scale?

    Ans. _____

20. The velocity equivalents of the Beaufort table are correct for any anemometer, regardless of exposure. (True or false?)

    Ans. _____

21. List all the steps necessary in measuring depth of snow on ground.

    Ans. _____

    _____

22. Name three steps in caring for the weighing-type precipitation gage during the winter.

    Ans.   *a.* _____

            b. _____

            c. _____

23. What clouds are most generally referred to as the high types?

    Ans. _____

    _____

24. What clouds are most generally referred to as the low types?

    Ans. _____

    _____

25. What clouds are most generally referred to as the middle types?

    Ans. _____

    _____

26. What types of clouds are most generally referred to as the vertical development types?

    Ans. _____

    _____

27. What is a stratus cloud called when its base rests on the ground?

    Ans. _____

    _____

28. What procedure is used for determining cloud directions?

    Ans. _____

29. If clouds are moving east, what direction is reported?

    Ans. _____

    _____

30. Which of the following is correct in determining sky conditions in relation to clouds present:

    a. Eighths_____   *c.* Sixths_____   *e.* Tenths_____

    b. Fourths_____   *d.* Fifths_____

31. A clinometer is an instrument used in _____

32. Define the word visibility in relation to weather observing and reporting.

    Ans. _____

    _____

33. Why would an airway beacon light not be a good object to use in estimating visibility at night?

    Ans. _____

    _____

34. The minimum thermometer reads 72.1. The correction card shows a correction of −0.3. What reading is recorded?

    Ans. _____

35. The maximum thermometer reads 72.1. The correction card shows a correction of −0.3. What reading is recorded?

    Ans. _____

36. The dry-bulb thermometer reads 39.0. The correction is −0.1. How should this reading be recorded?

    Ans. _____

37. Why is it necessary to use an alcohol thermometer in northern Canada?

    Ans. _____

38. Give three methods of determining true north.

    Ans. *a.* _____ *b.* _____ *c.* _____

39. Should a wind vane be oriented to magnetic north or true north for meteorological observations?

    Ans. _____

40. How are winds aloft determined?

    Ans. _____

41. Why do synoptic weather observing stations reduce station pressure to sea level?

    Ans. _____

    _____

42. On a day with cumulus clouds present, temperature 80° F., dew point 60° F., what height would the cloud bases be?

    Ans. _____

43. How are temperature, pressure, and humidity measured above the surface of the earth?

    Ans. _____

    _____

44. Is it possible for the dew point to have a higher value than the dry-bulb temperature?

    Ans. _____

45. How is visibility determined?

    Ans. _____

    _____

46. How is gustiness determined in wind observations?

    Ans. _____

47. Does a temperature correction have to be applied to the usual compensated aneroid barometer to obtain station pressure?

    Ans. _____

48. What are the differences between altocumulus and cirrocumulus clouds?

    Ans. _____

    _____

49. What instrument is used to measure cloud direction?

    Ans. _____

50. What are some of the reasons for establishing a weather observatory?

    Ans. _____

    _____

SURFACE WEATHER OBSERVATIONS
(LAND STATION)

Station _____

Lat. _____ Long. _____

Time entries on this form are _____ th meridian

To convert } { add
to G. C. T. } { subtract } _____ hours

Height of Barometer _____ Ft. (MSL)

DATE	TIME	BAROMETER			THERMOMETERS					SELF-REG. THER'S		PRECIP. (Inches)	WIND			LOW		MIDDLE		HIGH		WEATHER AND/OR OBSTRUCTIONS TO VISION	REMARKS	OBS. INIT.
		STATION (OBSERVED READING PLUS TOTAL COR.)	REDUCED TO SEA LEVEL	PRES. TEND.	NET 3-HR. CHANGE (Inches)	DRY	WET	DEW POINT	REL- A-TIVE HU-MID-ITY	VAPOR PRES-SURE	MAXI-MUM	MINI-MUM		DIREC-TION	SPEED (M.P.H.)	CHAR-ACTER AND SHIFTS	AMOUNT	AMT., TYPE, DIRECTION	HEIGHT (Yards or Ft.)	AMT., TYPE, DIRECTION	HEIGHT (Yards or Ft.)	AMT., TYPE, DIRECTION	HEIGHT (Yards or Ft.)			

CLOUDS

SURFACE WEATHER OBSERVATIONS
(LAND STATION)

Time entries on this form are _____ th meridian

To convert { add
to G. C. T. { subtract } _____ hours

Height of Barometer _____ Ft. (MSL)

Station _____
Lat. _____ Long. _____

DATE	TIME	BAROMETER			THERMOMETERS					SELF-REG. THER'S		PRECIP. (inches)	WIND			CLOUDS						WEATHER AND/OR OBSTRUCTIONS TO VISION	REMARKS	OBS. INIT.
		STATION (OBSERVED READING PLUS TOTAL COR.)	REDUCED TO SEA LEVEL	PRES. 3-HR. TEND.	NET 3-HR. CHANGE (inches)	DRY	WET	DEW POINT	RELATIVE HUMIDITY	VAPOR PRESSURE	MAXI-MUM	MINI-MUM		DIREC-TION	SPEED (m.p.h.)	CHARACTER AND SHIFTS	AMOUNT	LOW		MIDDLE		HIGH				
																		AMT., TYPE, DIRECTION	HEIGHT (hmds. or Ft.)	AMT., TYPE, DIRECTION	HEIGHT (hmds. or Ft.)	AMT., TYPE, DIRECTION	HEIGHT (hmds. or Ft.)			

SURFACE WEATHER OBSERVATIONS
(LAND STATION)

Time entries on this form are _____ th meridian

To convert { add / subtract } to G. C. T. _____ hours

Height of Barometer _____ Ft. (MSL)

Station _____
Lat. _____ Long. _____

DATE	TIME	BAROMETER			THERMOMETERS					SELF-REG. THER'S		PRECIP. (Inches)	WIND				CLOUDS						WEATHER AND/OR OBSTRUCTIONS TO VISION	REMARKS	OBS. INIT.
		STATION (OBSERVED READING PLUS TOTAL COR.)	REDUCED TO SEA LEVEL	PRES. TEND. / NET 3-HR. CHANGE (Inches)	DRY	WET	DEW POINT	RELA-TIVE HU-MID-ITY	VAPOR PRES-SURE	MAXI-MUM	MINI-MUM		DIREC-TION	SPEED (M.P.H.)	CHAR-ACTER AND SHIFTS	AMOUNT	LOW AMT., TYPE, DIRECTION	LOW HEIGHT (Hds. or Ft.)	MIDDLE AMT., TYPE, DIRECTION	MIDDLE HEIGHT (Hds. or Ft.)	HIGH AMT., TYPE, DIRECTION	HIGH HEIGHT (Hds. or Ft.)			

SURFACE WEATHER OBSERVATIONS
(LAND STATION)

Station _____

Lat. _____ Long. _____

Time entries on this form are _____ th meridian

To convert { add } _____ hours
to G. C. T. { subtract }

Height of Barometer _____ Ft. (MSL)

DATE	TIME	BAROMETER			THERMOMETERS					SELF-REG. THER'S		PRECIP. (Inches)	WIND			AMOUNT	CLOUDS									WEATHER AND/OR OBSTRUCTIONS TO VISION	REMARKS	OBS. INIT.
		STATION (OBSERVED READING PLUS TOTAL COR.)	REDUCED TO SEA LEVEL	PRES. TEND.	NET 3-HR. CHANGE (Inches)	DRY	WET	DEW POINT	RELA- TIVE HU- MID- ITY	VAPOR PRES- SURE	MAXI- MUM	MINI- MUM		DIREC- TION	SPEED (M. P. H.)	CHAR- ACTER AND SHIFTS		LOW		MIDDLE		HIGH						
																		AMT., TYPE, DIRECTION	HEIGHT (Hds. or Ft.)	AMT., TYPE, DIRECTION	HEIGHT (Hds. or Ft.)	AMT., TYPE, DIRECTION	HEIGHT (Hds. or Ft.)					

SURFACE WEATHER OBSERVATIONS

(LAND STATION)

Time entries on this form are _____ th meridian

To convert { add } _____ hours
to G. C. T. { subtract }

Height of Barometer _____ Ft. (MSL)

Station _____

Lat. _____ Long. _____

DATE	TIME	BAROMETER				THERMOMETERS			RELA-TIVE HU-MID-ITY	VAPOR PRES-SURE	SELF-REG. THER'S		WIND				CLOUDS						WEATHER AND/OR OBSTRUCTIONS TO VISION	REMARKS	OBS. INIT.
		STATION (OBSERVED READING PLUS TOTAL COR.)	REDUCED TO SEA LEVEL	PRES. 3-HR. TEND.	NET 3-HR. CHANGE (Inches)	DRY	WET	DEW POINT			MAXI-MUM	MINI-MUM	PRECIP. (Inches)	DIREC-TION	SPEED (M.P.H.)	CHAR-ACTER AND SHIFTS	AMOUNT	LOW		MIDDLE		HIGH				
																		AMT., TYPE, DIRECTION	HEIGHT (Hars, or Ft.)	AMT., TYPE, DIRECTION	HEIGHT (Hars, or Ft.)	AMT., TYPE, DIRECTION	HEIGHT (Hars, or Ft.)			

SURFACE WEATHER OBSERVATIONS
(LAND STATION)

Station _____
Lat. _____   Long. _____

Time entries on this form are _____ th meridian

To convert to G. C. T. { add / subtract } _____ hours

Height of Barometer _____ Ft. (MSL)

DATE	TIME	BAROMETER		NET 3-HR. TEND. CHANGE (INCH)	THERMOMETERS			RELA- TIVE HU- MID- ITY	VAPOR PRES- SURE	SELF-REG. THER'S		PRECIP. (INCHES)	WIND			AMOUNT	CLOUDS						WEATHER AND/OR OBSTRUCTIONS TO VISION	REMARKS	OBS. INIT.
		STATION (OBSERVED READING PLUS TOTAL COR.)	REDUCED TO SEA LEVEL	PRES. 3-HR. TEND. CHANGE (INCH)	DRY	WET	DEW POINT			MAXI- MUM	MINI- MUM		DIREC- TION	SPEED (M. P. H.)	CHAR- ACTER AND SHIFTS		LOW AMT. TYPE, DIRECTION	LOW HEIGHT (MEAS. OF FT.)	MIDDLE AMT. TYPE, DIRECTION	MIDDLE HEIGHT (MEAS. OF FT.)	HIGH AMT. TYPE, DIRECTION	HIGH HEIGHT (MEAS. OF FT.)			

SURFACE WEATHER OBSERVATIONS
(LAND STATION)

Time entries on this form are _____ th meridian

To convert { add } to G. C. T. { subtract } _____ hours

Height of Barometer _____ Ft. (MSL)

Station_____

Lat._____ Long._____

DATE	TIME	BAROMETER				THERMOMETERS					SELF-REG. THER'S		PRECIP. (Inches)	WIND				CLOUDS											WEATHER AND/OR OBSTRUCTIONS TO VISION	REMARKS	OBS. INIT.
		STATION (OBSERVED READING PLUS TOTAL COR.)	REDUCED TO SEA LEVEL	PRES. TEND.	NET 3-HR. CHANGE (Inches)	DRY	WET	DEW POINT	RELA-TIVE HU-MID-ITY	VAPOR PRES-SURE	MAXI-MUM	MINI-MUM		DIREC-TION	SPEED (M.P.H.)	CHAR-ACTER AND SHIFTS	AMOUNT	LOW			MIDDLE			HIGH					
																		AMT., TYPE, DIRECTION	HEIGHT (Mes. or Ft.)		AMT., TYPE, DIRECTION	HEIGHT (Mes. or Ft.)		AMT., TYPE, DIRECTION	HEIGHT (Mes. or Ft.)				

SURFACE WEATHER OBSERVATIONS
(LAND STATION)

Station _____
Lat. _____ Long. _____

Time entries on this form are _____ th meridian

To convert { add. } _____ hours
to G. C. T. { subtract }

Height of Barometer _____ Ft. (MSL)

DATE	TIME	BAROMETER			THERMOMETERS					SELF-REG. THER'S		PRECIP. (inches)	WIND				CLOUDS						WEATHER AND/OR OBSTRUCTIONS TO VISION	REMARKS	OBS. INT.
		STATION (OBSERVED READING PLUS TOTAL COR.)	REDUCED TO SEA LEVEL	NET 3-HR. TEND. CHANGE (inches)	PRES.-URE TEND.	DRY	WET	DEW POINT	RELA-TIVE HU-MID-ITY	VAPOR PRES-SURE	MAXI-MUM	MINI-MUM		DIREC-TION	SPEED (M.P.H.)	CHAR-ACTER AND SHIFTS	AMOUNT	LOW		MIDDLE		HIGH				
																		AMT., TYPE, DIRECTION	HEIGHT (HDS. OR FT.)	AMT., TYPE, DIRECTION	HEIGHT (HDS. OR FT.)	AMT., TYPE, DIRECTION	HEIGHT (HDS. OR FT.)			

SURFACE WEATHER OBSERVATIONS
(LAND STATION)

Time entries on this form are _____ th meridian

To convert } { add
to G. C. T. } { subtract } _____ hours

Height of Barometer _____ Ft. (MSL)

Station _____
Lat. _____ Long. _____

DATE	TIME	BAROMETER		THERMOMETERS						SELF-REG. THER'S		PRECIP. (inches)	WIND				CLOUDS								WEATHER AND/OR OBSTRUCTIONS TO VISION	REMARKS	OBS. INIT.
		STATION (OBSERVED READING PLUS TOTAL CORR.)	REDUCED TO SEA LEVEL	PRES. 3-HR. TEND. (inches)	NET 3-HR. CHANGE (inches)	DRY	WET	DEW POINT	RELA- TIVE HU- MID- ITY	VAPOR PRES- SURE	MAXI- MUM	MINI- MUM		DIREC- TION	SPEED (M.P.H.)	CHAR- ACTER AND SHIFTS	AMOUNT	LOW			MIDDLE			HIGH				
																		AMT., TYPE, DIRECTION	HEIGHT (Hdrs, or Ft.)		AMT., TYPE, DIRECTION	HEIGHT (Hdrs. or Ft.)		AMT., TYPE, DIRECTION	HEIGHT (Hdrs. or Ft.)			

SURFACE WEATHER OBSERVATIONS
(LAND STATION)

Station _____

Lat. _____ Long. _____

Time entries on this form are ___ th meridian

To convert to G. C. T. { add / subtract } ___ hours

Height of Barometer ___ Ft. (M.S.L.)

DATE	TIME	BAROMETER			THERMOMETERS				RELATIVE HUMIDITY	VAPOR PRESSURE	SELF-REG. THERS		PRECIP. (Inches)	WIND			AMOUNT	CLOUDS							WEATHER AND/OR OBSTRUCTIONS TO VISION	REMARKS	OBS. INITS.
		STATION (OBSERVED READING PLUS TOTAL CORR.)	REDUCED TO SEA LEVEL	NET 3-HR. PRES. TEND. CHANGE (Inches)	DRY	WET	DEW POINT				MAXI-MUM	MINI-MUM		DIREC-TION	SPEED (M.P.H.)	CHAR-ACTER AND SHIFTS		LOW			MIDDLE			HIGH				
																		AMT., TYPE, DIRECTION	HEIGHT (MBS. OR FT.)		AMT., TYPE, DIRECTION	HEIGHT (MBS. OR FT.)		AMT., TYPE, DIRECTION	HEIGHT (MBS. OR FT.)			

SURFACE WEATHER OBSERVATIONS
(LAND STATION)

Time entries on this form are _____ th meridian

To convert { add / subtract } to G. C. T. _____ hours

Height of Barometer _____ Ft. (MSL)

Station _____

Lat. _____ Long. _____

DATE	TIME	BAROMETER			THERMOMETERS		DEW POINT	RELATIVE HUMIDITY	VAPOR PRESSURE	SELF-REG. THER'S		PRECIP. (INCHES)	WIND			AMOUNT	CLOUDS									WEATHER AND/OR OBSTRUCTIONS TO VISION	REMARKS	OBS. INIT.
		STATION (OBSERVED READING PLUS TOTAL COR.)	REDUCED TO SEA LEVEL	NET 3-HR. TEND. CHANGE (INCHES)	DRY	WET				MAXI- MUM	MINI- MUM		DIREC- TION	SPEED (M. P. H.)	CHAR- ACTER AND SHIFTS		LOW			MIDDLE			HIGH					
																	AMT., TYPE, DIRECTION	HEIGHT (HNDS. OF FT.)		AMT., TYPE, DIRECTION	HEIGHT (HNDS. OF FT.)		AMT., TYPE, DIRECTION	HEIGHT (HNDS. OF FT.)				

SURFACE WEATHER OBSERVATIONS
(LAND STATION)

Time entries on this form are _____ th meridian

To convert { add
to G. C. T. { subtract } _____ hours

Height of Barometer _____ Ft. (MSL)

Station _____
Lat. _____ Long. _____

DATE	TIME	BAROMETER			THERMOMETERS					SELF-REG. THER'S		PRECIP. (inches)	WIND			AMOUNT	CLOUDS									WEATHER AND/OR OBSTRUCTIONS TO VISION	REMARKS	OBS. INIT.
		STATION (OBSERVED READING PLUS TOTAL CORR.)	REDUCED TO SEA LEVEL	PRES. TEND.	NET 3-HR. CHANGE (inches)	DRY	WET	DEW POINT	RELA-TIVE HU-MID-ITY	VAPOR PRES-SURE	MAXI-MUM	MINI-MUM		DIREC-TION	SPEED (M.P.H.)	CHAR-ACTER AND SHIFTS		LOW		MIDDLE		HIGH					
																		AMT. TYPE, DIRECTION	HEIGHT (Hds. or Ft.)	AMT. TYPE, DIRECTION	HEIGHT (Hds. or Ft.)	AMT. TYPE, DIRECTION	HEIGHT (Hds. or Ft.)				

SURFACE WEATHER OBSERVATIONS
(LAND STATION)

Time entries on this form are _____ th meridian

To convert to G. C. T. { add / subtract } _____ hours

Height of Barometer _____ Ft. (MSL)

Station _____ Lat. _____ Long. _____

DATE	TIME	BAROMETER		PRES. TEND.	NET 3-HR. CHANGE (Inches)	THERMOMETERS		DEW POINT	RELA- TIVE HU- MID- ITY	VAPOR PRES- SURE	SELF-REG. THER'S		PRECIP. (Inches)	WIND				CLOUDS									WEATHER AND/OR OBSTRUCTIONS TO VISION	REMARKS	OBS. INT.
		STATION (OBSERVED READING PLUS TOTAL COR.)	REDUCED TO SEA LEVEL			DRY	WET				MAXI- MUM	MINI- MUM		DIREC- TION	SPEED (M.P.H.)	CHAR- ACTER AND SHIFTS	AMOUNT	LOW			MIDDLE			HIGH					
																		AMT. TYPE, DIRECTION	HEIGHT (MRS. OR FT.)		AMT. TYPE, DIRECTION	HEIGHT (MRS. OR FT.)		AMT. TYPE, DIRECTION	HEIGHT (MRS. OR FT.)				

SURFACE WEATHER OBSERVATIONS
(LAND STATION)

Station _____
Lat. _____
Long. _____

Time entries on this form are _____ th meridian

To convert { add } to G. C. T. { subtract } _____ hours

Height of Barometer _____ Ft. (MSL)

| DATE | TIME | BAROMETER | | | THERMOMETERS | | | | | SELF-REG. THER'S | | PRECIP. (INCHES) | WIND | | | | CLOUDS | | | WEATHER AND/OR OBSTRUCTIONS TO VISION | REMARKS | OBS. INIT. |
| | | STATION (OBSERVED READING PLUS TOTAL CORR.) | REDUCED TO SEA LEVEL | PRES. TEND. | NET 3-HR. CHANGE (INCHES) | DRY | WET | DEW POINT | RELATIVE HUMIDITY | VAPOR PRESSURE | MAXIMUM | MINIMUM | | DIRECTION | SPEED (M.P.H.) | CHARACTER AND SHIFTS | AMOUNT | LOW — AMT., TYPE, DIRECTION / HEIGHT (HDS. OR FT.) | MIDDLE — AMT., TYPE, DIRECTION / HEIGHT (HDS. OR FT.) | HIGH — AMT., TYPE, DIRECTION / HEIGHT (HDS. OR FT.) | | | |

SURFACE WEATHER OBSERVATIONS
(LAND STATION)

Time entries on this form are _____ th meridian

To convert to G. C. T. { add / subtract } _____ hours

Height of Barometer _____ Ft. (MSL)

Station _____

Lat. _____ Long. _____

DATE	TIME	BAROMETER				THERMOMETERS				VAPOR PRES-SURE	SELF-REG. THER'S		PRECIP. (Inches)	WIND			AMOUNT	CLOUDS												WEATHER AND/OR OBSTRUCTIONS TO VISION	REMARKS	OBS. INIT.
		STATION (OBSERVED READING PLUS TOTAL COR.)	REDUCED TO SEA LEVEL	PRES. TEND.	NET 3-HR. CHANGE (Inches)	DRY	WET	DEW POINT	RELA-TIVE HU-MID-ITY		MAXI-MUM	MINI-MUM		DIREC-TION	SPEED (M.P.H.)	CHAR-ACTER AND SHIFTS		LOW				MIDDLE				HIGH						
																		AMT., TYPE, DIRECTION		HEIGHT (Mets. or Ft.)		AMT., TYPE, DIRECTION		HEIGHT (Mets. or Ft.)		AMT., TYPE, DIRECTION		HEIGHT (Mets. or Ft.)				

SURFACE WEATHER OBSERVATIONS
(LAND STATION)

Time entries on this form are _____ th meridian

To convert { add } _____ hours
to G. C. T. { subtract }

Height of Barometer _____ Ft. (MSL)

Station _____
Lat. _____ Long. _____

DATE	TIME	BAROMETER		PRES. TEND.	NET 3-HR. CHANGE (INCHES)	THERMOMETERS		DEW POINT	RELA- TIVE HU- MID- ITY	VAPOR PRES- SURE	SELF-REG. THER'S		PRECIP. (INCHES)	WIND			AMOUNT	CLOUDS						WEATHER AND/OR OBSTRUCTIONS TO VISION	REMARKS	OBS. INIT.
		STATION (OBSERVED READING PLUS TOTAL COR.)	REDUCED TO SEA LEVEL			DRY	WET				MAXI- MUM	MINI- MUM		DIREC- TION	SPEED (M.P.H.)	CHAR- ACTER AND SHIFTS		LOW		MIDDLE		HIGH				
																		AMT., TYPE, DIRECTION	HEIGHT (HND., OR FT.)	AMT., TYPE, DIRECTION	HEIGHT (HND., OR FT.)	AMT., TYPE, DIRECTION	HEIGHT (HND., OR FT.)			

SURFACE WEATHER OBSERVATIONS
(LAND STATION)

Time entries on this form are _____ th meridian

To convert { add } _____ hours
to G. C. T. { subtract }

Height of Barometer _____ Ft. (MSL)

Station _____
Lat. _____ Long. _____

DATE	TIME	BAROMETER			THERMOMETERS					SELF-REG. THER'S		PRECIP. (Inches)	WIND				CLOUDS							WEATHER AND/OR OBSTRUCTIONS TO VISION	REMARKS	OBS. INIT.
		STATION (OBSERVED READING PLUS TOTAL COR.)	REDUCED TO SEA LEVEL	PRES. TEN-DEN.	NET 3-HR. CHANGE (Inches)	DRY	WET	DEW POINT	RELATIVE HUMIDITY	VAPOR PRES-SURE	MAXI-MUM	MINI-MUM		DIREC-TION	SPEED (M. P. H.)	CHAR-ACTER AND SHIFTS	AMOUNT	LOW		MIDDLE		HIGH				
																		AMT., TYPE, DIRECTION	HEIGHT (Mtrs. or Ft.)	AMT., TYPE, DIRECTION	HEIGHT (Mtrs. or Ft.)	AMT., TYPE, DIRECTION	HEIGHT (Mtrs. or Ft.)			

SURFACE WEATHER OBSERVATIONS
(LAND STATION)

Time entries on this form are _____ th meridian

To convert { add } to G. C. T. { subtract } _____ hours

Height of Barometer _____ Ft. (MSL)

Station _____

Lat. _____ Long. _____

DATE	TIME	BAROMETER				THERMOMETERS					SELF-REG. THER'S		PRECIP. (Inches)	WIND				CLOUDS												WEATHER AND/OR OBSTRUCTIONS TO VISION	REMARKS	OBS. INIT.
		STATION (OBSERVED READING PLUS TOTAL COR.)	REDUCED TO SEA LEVEL	PRES. TEND.	NET 3-HR. CHANGE (Inches)	DRY	WET	DEW POINT	RELATIVE HUMIDITY	VAPOR PRESSURE	MAXIMUM	MINIMUM		DIRECTION	SPEED (M.P.H.)	CHARACTER AND SHIFTS	AMOUNT	LOW				MIDDLE				HIGH						
																		AMT., TYPE, DIRECTION	HEIGHT (Mets. or Ft.)			AMT., TYPE, DIRECTION	HEIGHT (Mets. or Ft.)			AMT., TYPE, DIRECTION	HEIGHT (Mets. or Ft.)					

SURFACE WEATHER OBSERVATIONS
(LAND STATION)

Time entries on this form are _____ th meridian

To convert { add.
to G. C. T. { subtract } _____ hours

Height of Barometer _____ Ft. (MSL)

Station _____
Lat. _____ Long. _____

DATE	TIME	BAROMETER			THERMOMETERS					SELF-REG. THER'S		PRECIP. (INCHES)	WIND			AMOUNT	CLOUDS								WEATHER AND/OR OBSTRUCTIONS TO VISION	REMARKS	OBS. INIT.
		STATION (OBSERVED READING PLUS TOTAL COR.)	REDUCED TO SEA LEVEL	PRES. TEND.	NET 3-HR. CHANGE (INCHES)	DRY	WET	DEW POINT	RELATIVE HUMIDITY	VAPOR PRESSURE	MAXIMUM	MINIMUM		DIRECTION	SPEED (M.P.H.)	CHARACTER AND SHIFTS		LOW			MIDDLE			HIGH				
																		AMT., TYPE, DIRECTION	HEIGHT (HNDS. OR FT.)		AMT., TYPE, DIRECTION	HEIGHT (HNDS. OR FT.)		AMT., TYPE, DIRECTION	HEIGHT (HNDS. OR FT.)				

SURFACE WEATHER OBSERVATIONS
(LAND STATION)

Station _____
Lat. _____
Long. _____

Time entries on this form are _____ th meridian

To convert to G. C. T. { add / subtract } _____ hours

Height of Barometer _____ Ft. (MSL)

DATE	TIME	BAROMETER			THERMOMETERS			RELA-TIVE HU-MID-ITY	VAPOR PRES-SURE	SELF-REG. THER'S		PRECIP. (inches)	WIND			AMOUNT	CLOUDS							WEATHER AND/OR OBSTRUCTIONS TO VISION	REMARKS	OBS. INIT.
		STATION (OBSERVED READING PLUS TOTAL COR.)	REDUCED TO SEA LEVEL	PRES. TEND.	NET 3-HR. CHANGE (inches)	DRY	WET	DEW POINT			MAXI-MUM	MINI-MUM		DIREC-TION	SPEED (m.p.h.)	CHAR-ACTER AND SHIFTS		LOW			MIDDLE			HIGH			
																		AMT., TYPE, DIRECTION	HEIGHT (hds. or Ft.)		AMT., TYPE, DIRECTION	HEIGHT (hds. or Ft.)		AMT., TYPE, DIRECTION	HEIGHT (hds. or Ft.)	

index